MANAGING AUTOMOTIVE BUSINESSES

Strategic Planning, Personnel, and Finances

MANAGING AUTOMOTIVE BUSINESSES

Strategic Planning, Personnel, and Finances

Ronald A. Garner, Ph.D.
Pennsylvania College of Technology
Williamsport, Pennsylvania

C. William Garner, D.Ed.
Rutgers University
New Brunswick, New Jersey

THOMSON
DELMAR LEARNING Australia • Canada • Mexico • Singapore • Spain • United Kingdom • United States

Managing Automotive Businesses: Strategic Planning, Personnel, and Finanaces
Garner - Garner

Vice President, Technology and Trades SBU:
Alar Elken

Editorial Director:
Sandy Clark

Senior Acquisitions Editor:
David Boelio

Development Editor:
Sharon Chambliss

Marketing Director:
David Garza

Channel Manager:
William Lawrenson

Marketing Coordinator:
Mark Pierro

Production Director:
Mary Ellen Black

Production Manager:
Andrew Crouth

Production Editor:
Jennifer Hanley

Art/Design Coordinator:
Nicole Stagg

Technology Project Manager:
Kevin Smith

Editorial Assistant:
Andrea Domkowski

Printed in Canada
1 2 3 4 5 XX 07 06 05

For more information contact
Thomson Delmar Learning
Executive Woods
5 Maxwell Drive, PO Box 8007,
Clifton Park, NY 12065-8007
Or find us on the World Wide Web at
www.delmarlearning.com

Library of Congress Cataloging-in-Publication Data
Garner, Ron, 1966–
Managing automotive businesses : strategic planning, personnel, and finances / Ron Garner, C. William Garner.
p. cm.
ISBN 1-4018-9896-3
1. Service stations—Management. I. Garner, C. William. II. Title.
TL153.G37 2006
629.28′6—dc22
2005021132

NOTICE TO THE READER

Publisher does not warrant or guarantee any of the products described herein or perform any independent analysis in connection with any of the product information contained herein. Publisher does not assume, and expressly disclaims, any obligation to obtain and include information other than that provided to it by the manufacturer.

The reader is expressly warned to consider and adopt all safety precautions that might be indicated by the activities herein and to avoid all potential hazards. By following the instructions contained herein, the reader willingly assumes all risks in connection with such instructions.

The publisher makes no representation or warranties of any kind, including but not limited to, the warranties of fitness for particular purpose or merchantability, nor are any such representations implied with respect to the material set forth herein, and the publisher takes no responsibility with respect to such material. The publisher shall not be liable for any special, consequential, or exemplary damages resulting, in whole or part, from the readers' use of, or reliance upon, this material.

CONTENTS

PREFACE

This book was written for people who wish to be employed as a service manager in an automotive facility. The authors assume that this person already has an in-depth knowledge of automobile systems as well as the job duties of a service consultant. The purpose of the book is to introduce basic management principles, personnel management, and the financial operations of automobile service facilities as they relate to the position of automotive service manager.

Recognizing that not all automobile service facilities are the same, this book points out the differences among them. This is critical because the variations in management structure, ownership, and strategic business plans influence the tasks and responsibilities assigned to a service manager. In addition, because the type of facility, such as small independent service facilities, system specific facilities, franchises, public/private fleets, and large dealerships, determine the service sold, the organizational structures and service manager's jobs vary from facility to facility. Consequently, this book attempts to present the differences in the position of the service manager and how their responsibilities and authority are altered in each type of facility.

Although the business objective of every company is to make a profit, what a facility does to earn a profit and how it attempts to do it differs regardless of the type of facility it is, form of ownership, or management structure. Comprehending how these variations affect a service manager's position is critical to his/her success. Therefore, the text covers all personnel tasks that may be performed by a service manager plus the financial operations of service facilities. Recognizing which of these tasks and responsibilities are, or are not, assigned is important to a service manager as well as individuals who are seeking such a position.

The authors were extremely careful to introduce and apply consistently throughout the book selected terms, processes, and formats used in business. Learning to use these terms in the proper context is important to a service manager when communicating with business professionals, such as an accountant, banker, consultant, employment agency representative insurance broker, vendor, and so on. In the text, the authors use the word *service* to represent a facility that provides automobile maintenance services or diagnostic and repair services. A service manager, of course, may work in a facility that offers both or either of these services. Making these distinctions clear is important to business relationships. In addition, regardless of the services offered, the importance of employee and business performances is a key interest when discussing business. Furthermore, in most cases, the owners of a facility have a clear-cut idea of what has to be done in their facility, and they will depend on a service manager to assist them in making their plan a reality. From the authors' experiences, the dependence on a service manager for advice is the growing trend and this is where some service managers have limitations or feel uncomfortable. Thus, this book emphasizes that managers must focus on performance, both in terms of personnel and financial performances, while at the same time keeping an eye on the bottom line (profit).

Book Overview

The first part of the text (Chapters 1–4) covers basic management principles. The primary topics covered in these chapters include the creation and application of management structures; types of ownerships and service facilities and their management structure, the position of the service manager in different structures, the purpose and elements of a strategic business plan, environmental effects on a business plan; and formatted systems.

The topic for Chapters 5–8 is personnel management. These chapters describe the tasks and duties of a service manager in the recruitment, selection, induction, development, and evaluation of employees. The chapters also cover interview techniques, formative and summative evaluations, the shaping process, performance objectives, and procedures for the termination or recognition of employees.

The third and fourth parts of the book (Chapters 9–15) focus on the financial operations of a service facility. Beginning with the presentation of basic business practices and financial statements used by managers, the chapters cover payroll and markups as well as the analysis of financial data.

Acknowledgments

The authors wish to recognize and express their appreciation to the following reviewers for the time they spent examing our work and offering their insights and helpful suggestions: Randall Peter Indiana State University and Mike Behrmann Southern Ilinois University.

From the Authors

The content in this book is based on our experiences as owners/operators of an automotive service facility; our studies of automotive technology, business, and vocational education; and our work as teachers/trainers and consultants. As owners/operators, we served as the service manager and sometimes the service consultant, cashier, manager, technician, and even the custodian. Starting an automotive service facility in an empty building with two old lifts and a compressor, we gained invaluable insights into the trials and tribulations associated with creating a business from the ground up. We hope that you will benefit from the knowledge we gained from our experiences.

Ron and Bill Garner

About the Authors

Ron Garner, Ph.D. is an Associate Professor of Automotive Technology in the Division of Transportation Technology at the Pennsylvania College of Technology, an affiliate of Pennsylvania State University. In that capacity he oversees a baccalaureate program in Automotive Technology Management, teaches a wide range of upper level technology courses, teaches Enhanced Pennsylvania Emission Certification, directs baccalaureate student final projects, and consults with organizations in business and nonprofit sectors.

He completed an A.A.S. degree at Ford ASSET automotive technology from Lehigh County Community College, a B.S. in Vocational Education, and an M.S. in Vocational Education with an emphasis in School Administration and Leadership from Pennsylvania State University. He earned his doctorate in Workforce Education at Pennsylvania State University with an emphasis in Industrial Training and Post Secondary School Administration. An A.S.E. certified Master Automobile Technician with the Advanced Engine Performance and Compressed Natural Gas Certification, he has been a technician as well as owned and managed automobile repair facilities.

C. William Garner, D.Ed. served as an airborne sonar technician in the U.S. Navy from 1959 to 1963. He earned a bachelor's degree in business education, a master's degree in higher education administration, and a doctorate in vocational education at Pennsylvania State University. He then took an appointment with Southern Illinois University at Carbondale as an Assistant Professor of Occupational Education and Site Administrator at March Air Force Base in California. His next appointment was with the University of Louisville as an Assistant Professor of Vocational Education and Coordinator for Educational Programs at Fort Knox.

In 1978 he received an appointment as an Associate Professor of Vocational Education at Rutgers University. As a professor at Rutgers he has served as Chair of the Department of Urban Education, Chair of the Graduate Department of Vocational Education, Executive Director of the Vocational Education Resource Center, and Acting Dean. Currently, Dr. Garner is an Associate Professor of Education Administration at Rutgers University and has written books on government accounting and school finance. In addition, he and his son, Ron, owned and operated an auto repair business.

PART I

INTRODUCTION TO STRATEGIC PLANNING

CHAPTER 1

THE SERVICE MANAGER AND THE BUSINESS STRUCTURE

LEARNING OBJECTIVES

Upon reading this chapter, students should be able to:

- *Prepare a diagram for a management structure in an automotive service facility.*
- *Explain the difference between line and staff positions.*
- *Describe how communication flows in a business organization.*
- *Explain the difference between a job description and job authority.*
- *Describe the relationship between an organizational structure and chain of command.*
- *Diagram a flat span of control and explain why it can create problems.*

Introduction

A service manager at an automotive service facility occupies a pivotal position that requires him or her to oversee and provide direction for all personnel and financial activities. To carry out these and other responsibilities, a service manager must have experience as a service consultant and knowledge of the services performed on the automobiles by the facility. To meet the expectations assigned to the job, the service manager must be able to plan, organize, lead, and control a service facility's operations. The service manager also ensures that each customer's service request is properly performed, that the customer is satisfied, and that a profit is generated on the job. However, the service manager always must make sure that the directives of the owners are followed and that company policies are enforced as intended.

Alpha Motors (a fictitious name) provides an example of a service manager's duty to provide oversight and direction of personnel and financial activities. Assume this facility must have a 40 percent gross profit (income less technician's labor costs) to have enough money to pay expenses (rent, manager's salary, overhead, etc.) and generate a profit so that the owner can pay his business loans or receive a return on his investment. In addition, assume that the owner's accountant calculated that a labor rate of $51.96 per hour is required to generate the 40 percent gross profit, given the volume of customers (number of automobiles) serviced at the facility each month. Also assume that one of the service consultants charged a labor rate of $35 per hour because a customer gave him free baseball tickets.

In the review of the daily financial activities of Alpha Motors, the service manager discovered the change in labor rate and determined the reason it occurred. Now the service manager has an obligation to the owners to discipline the service consultant. The disciplinary action should first require the service consultant to make up the difference between the labor that should have been charged and the amount collected. In addition, the service manager should provide some verbal counseling, a written reprimand, or a suspension with or without pay. Depending on the circumstances, the service manager may even find it necessary to terminate the service consultant because, after all, the person may have committed a fraudulent act and broke the trust held with the owner.

The point is that the service manager serves (is employed) at the pleasure of the owners. This means that the owners place considerable trust in the service manager and, in turn, the manager is obligated to protect the owners' interests in order to earn their trust. If the manager cannot or will not place this trust at the highest priority, then that person must be replaced by someone who can and will. The actions of the manager that most affect the owners' trust and interests (risk, profit, customer satisfaction, and business growth) include personnel and financial management, which are the two main topics of this book.

To begin the examination of the service manager's personnel and financial responsibilities to the owners, he or she must recognize how the service facility and the position of the service manager fits into the business: specifically, what is expected of the service operations. To whom does the service manager report? Who reports to them? Who is their boss's boss? In addition, what communication channels and practices are to be followed when receiving and transmitting information? When can people in positions above or below the service manager make direct contact with each other and when should they go through the service manager? How much authority do they have and is this information written down in a policy manual? The answers to these and other related questions are provided in this introductory chapter.

Basic Diagrams for a Management Structure

A diagram of the business's management structure should indicate the position of the service manager in a service facility. This diagram is often referred to as an organizational or managerial chart because it presents the chain of command (the positions of the owners and managers and their relationships).

Figure 1-1 presents a basic example of an organizational chart that shows the relationship between three managers in a business that is a service facility (as opposed to a service facility in an automotive dealership, which will be discussed later). In this figure, Manager B (service manager) and Manager C (parts manager) have equal authority and report to Manager A (service director). Therefore, Manager A, the service director, is the supervisor of both Managers B (service manager) and C (parts manager). This means A evaluates the performances of B and C who, in turn, supervise the people working under them.

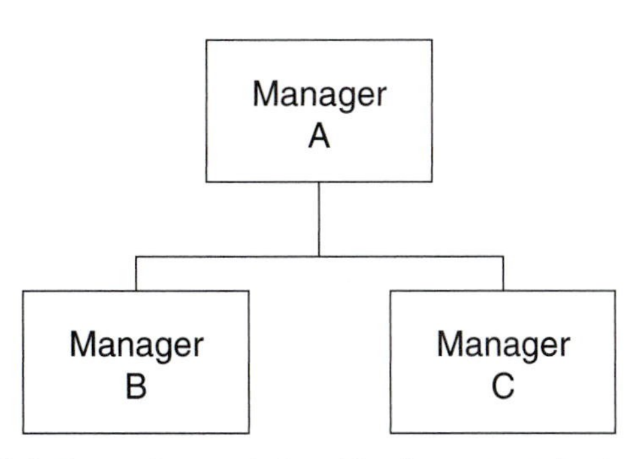

FIGURE 1-1 Supervisory relationships in an organizational chart.

Line versus Staff Managers

An important feature to be recognized in a flow diagram is the difference between **line** and **staff managers**. Specifically, a line manager has the authority to make company decisions and set company policy, such as a change in the labor rate to be charged to a special account, and typically, staff managers do not. Rather, staff managers are hired to support a line manager and, as a result, are delegated duties by the line manager. It is not unusual for staff managers to have employees work for them and, therefore, serve a line manager within a unit. This allows them to have the authority to make some decisions (depending on the line manager) within their unit or work group. For example, a service manager would be a line manager for the service department, while an assistant service manager would have the position of a staff manager with job duties similar to a service consultant. In this arrangement, the assistant service manager might be in charge of a group of four to six technicians who make up a work group or unit. The ability to delegate authority to the assistant manager would depend upon the service manager, who is in the line position.

Figure 1-2 shows a **flow diagram** for line and staff managers. Note that the service director is a line manager and in a position located below the owners. The assistant to the service director is in a staff position and shown in the diagram to the side and not below the service director.

Figure 1-2 also includes two department managers who report to the service director. Specifically, Manager A is the service manager and Manager B is the parts manager. This indicates that the service director is responsible for their supervision and performance evaluations. Naturally, the service manager (Manager A) and the parts manager (Manager B) both have employees in their departments and would be considered line managers for their departments and the employees they manage. This means

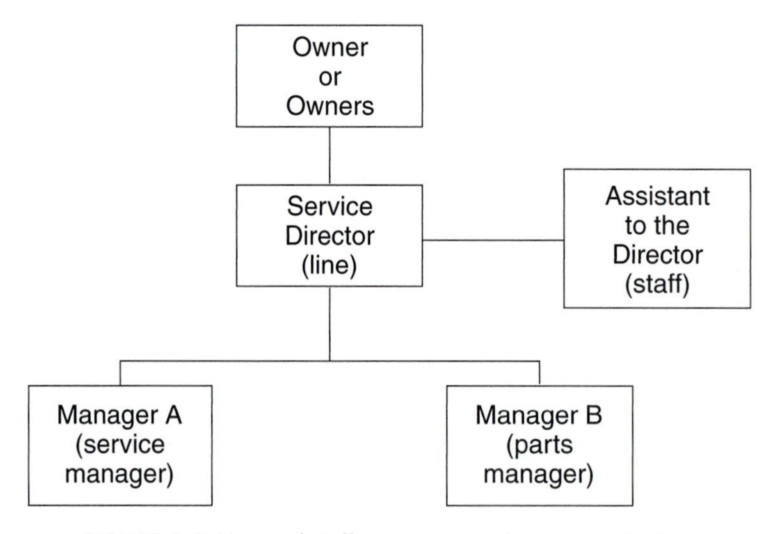

FIGURE 1-2 Line and staff management in an organization.

they would be responsible for the supervision and performance evaluations of their employees.

It is important to recognize what happens when the line manager is absent. As shown in Figure 1-2, either Manager A or B should have the ability and authority to act as the service director if the director is absent. Although some organizations permit the assistant to the service director to perform the daily duties of the service director when he/she is absent, the authority to make any major decisions rests with the owners (above the service director) or, in some cases, with one or both of the line managers located at the level below the service director.

An error is often made when a staff manager, such as the assistant to the service director, assumes the responsibilities and authority of the line manager, such as a service director, and makes major decisions in the person's absence. The reason this is a mistake is because the staff manager is typically a specialist, such as a data collection expert, warranty clerk, cashier, or bookkeeper. Consequently, staff manager specialists typically do not have the background, experience, knowledge, or expertise required of line managers.

As a result, in the diagram presented in Figure 1-2, when the service director is absent, Manager A or B should serve as a substitute when complex problems come up. With respect to major actions to be taken, Manager A or B might make decisions in the service director's absence with the approval of the owners, such as finalization of a fleet contract. In other words, although a staff manager can fulfill the routine duties of the service director, care must be taken to ensure the appointment of a qualified substitute (especially in the case of a long-term absence). The selection of the substitute should be based on several factors, including work experience in the business (i.e., seniority), the qualifications (i.e., knowledge of automobile systems, computer operation, among others) needed to handle complex problems, and a reputation for making correct decisions.

Decision Making Authority and the Chain of Command

The authority assigned to a position should not be confused with job duties. Job duties make up the tasks that a person performs in a position. Authority assigned to a position includes the power to give commands, enforce directives, make decisions, and take actions. Job duties are the same for everyone who occupies a position. However, the authority assigned to a position usually depends upon the person in the position and may change (more or less authority is given) from person to person. For example, as a service manager gains experience and the confidence and trust of the owners, he or she may be given greater authority to plan, organize, lead, and control. For instance, the service manager may be given the authority to supervise and/or direct other managers (such as the parts manager), hire employees in the department at will, suspend or terminate employees, make financial decisions (raise the labor rate), promote the business (prepare advertisements), or assume owner responsibilities (contact the company attorney when a serious problem arises).

New managers should—but often do not—have a clear understanding of the authority assigned to them by the owners or upper management. They might assume they have the same authority assigned to the person they replaced, but often they do not. Learning by trial and error how much authority a manager has been given is dangerous. New managers must be very clear about how much authority they have. For example, as shown in Figure 1-2, the service director reports to the owner. In one business, the director may be expected to contact the company attorney or place advertisements and then report the actions to the owner or owners. In other businesses, the director may be expected to make a recommendation to the owners, who would then contact the company attorney or decide whether or not to run an advertisement. Obviously, knowing the reach of one's authority to make a decision is critical.

Supervisory authority is presented in an organizational chart, such as the one shown in Figure 1-3. In this figure, arrows were inserted into the chart to indicate the direction in which authority flows. This chain of command indicates the direction in which power runs. Staff managers (service consultants, assistant service managers, assistant to the service director, lead technicians, and shop foreman) are not shown in a chain of command because they do not possess authority to make financially related decisions or hire and reprimand employees they supervise. This authority rests with the line manager, such as the service manager.

How much authority is provided to the service director in Figure 1-3 rests with the owners. Likewise, with the assistance of the owner, the

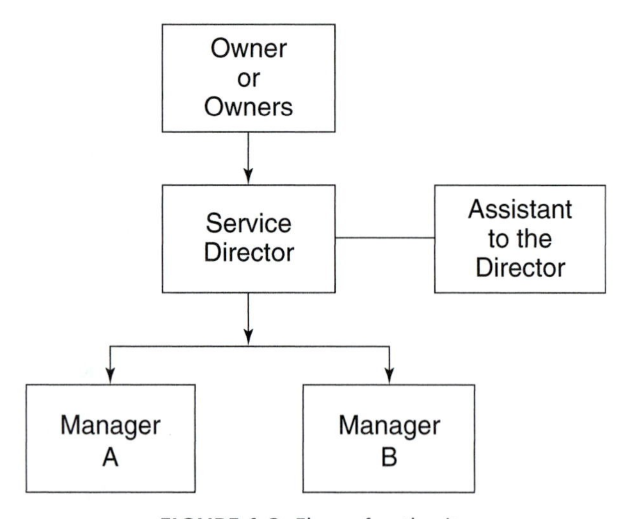

FIGURE 1-3 Flow of authority.

service director will often determine how much authority is delegated to Managers A and B. In some cases, one manager, such as the service manager, may be given greater authority than another; however, delegated authority must be constant and should not change from day to day. In cases in which assigned authority is assumed by a superior or owner unannounced or when it changes from day to day, the effectiveness of a manager is seriously compromised.

Because the staff manager shown in Figure 1-3 serves as the assistant to the service director, the service director assigns the job duties to this position. In addition, because the assistant to the director (staff manager position) is not a line position, the service director should not shift certain duties (such as hiring, disciplinary actions, or contract settlements) to this staff position. Furthermore, the service director (line manager) must remember that he or she is responsible for the actions and decisions of the staff manager. In other words, although tasks assigned to the line manager by the owner (or a superior line manager) may be shifted to a staff manager, the *responsibility* of the line manager to the owners (or superior manager) for actions taken by the staff manager cannot be shifted to the staff manager.

Consequently, a staff manager should only be given authority that is specific to the job assigned, as the line manager (service director) is ultimately responsible to the owners. For example, if a warranty claim to be submitted has an error, the staff manager may or may not have the authority to correct the error. If the error is a result of a number that cannot be read clearly because it was copied incorrectly, the staff manager should change the number and submit the documents without notifying the line manager. However, if the mileage shown is less than the number shown in the last repair invoice for the automobile, then the staff manager should not have the authority to change the number. Rather, the error should be reported to the line manager, who should look into the discrepancy. The difference, of course, may be because a mistake was made when the mileage was copied on to the form; however, it could suggest that someone did not follow procedures or even that someone committed an illegal act.

The line manager must assume responsibility for a discrepancy. This is important because the warranty company could interpret an incorrect mileage as a fraudulent action. This could lead to an investigation, charges, and penalties as serious as prison time for the line manager and even the owners, the violation of a franchise license agreement, or fines and/or punitive fees. A staff manager would not be expected to be aware of these possibilities, but the service manager must be.

Therefore, authority and the delegation of duties must be clearly assigned to the right people. Businesses have lost contracts, jobs have been lost, and federal prisons have inmates who were professionals (owners, executives, managers, accountants, bookkeepers) who were convicted of crimes because of their negligence related to the illegal actions of a subordinate. This means they held the authority to oversee the process, allowed a subordinate to make unsupervised decisions, and chose to ignore their responsibilities to the owners or stockholders.

Span of Control

The number of employees a manager personally supervises is known as **span of control**. For example, if the service director (Manager A) supervises Managers B and C, the span of control for Manager A is over the other two line managers plus the staff manager. A problem in some businesses is that, in their expansion, the span of control for the managers exceeds the number of people they can properly control. When the span of control extends beyond what a manager can manage (referred to as flat managerial structure), then the owners must restructure the organizational diagram and create a new structure. The revised structure may include additional managerial positions to transfer responsibilities to or subordinate managers may be hired to help with supervision. A staff manager may even be hired to assume additional assignments.

Most independent automotive service facilities start out with owners and workers. As the business grows and additional technicians are hired, the owners will usually hire people to assist in the running of the business, such as a receptionist, bookkeeper, or cashier. When the business continues to grow and more technicians are hired, the owners typically hire a manager who also serves as the service consultant. As more technicians continue to be added to the payroll, a full-time service consultant and possibly a parts specialist may be hired. If the business still continues to grow, the service manager cannot physically provide the direct supervision of the services, personnel, and financial activities. Consequently, additional people must be hired, and the service consultants may become assistant service managers with supervisory responsibilities over groups of technicians. Often, a manager may not want to give up this direct hands-on supervision even though the job requires 12-hour days over a 6-day workweek. When personnel are added, owners must change the structure or be faced with ineffective and inefficient performances that can lead to a decline in profits.

An example of a flat structure is shown in Figure 1-4, in which a Service Manager (SM-A) is responsible for six technicians (T-#1 to T-#6), two service consultants (SC-#1 and SC-#2), and a cashier and parts clerk. Because six technicians, two service consultants, and two staff employees (ten people) are too many for the Service Manager (SM-A) to supervise, the owners must modify the organizational structure of their business.

In the case of the structure shown in Figure 1-4, a modified structure could be prepared, as shown in Figure 1-5. In this diagram the parts clerk becomes a parts manager and the Service Manager oversees and supervises the Service Consultants (SC-#1 and SC-#2). Each service consultant is then responsible for three technicians. At the same time, the cashier has been made the Business Manager and supervises a cashier and accounts

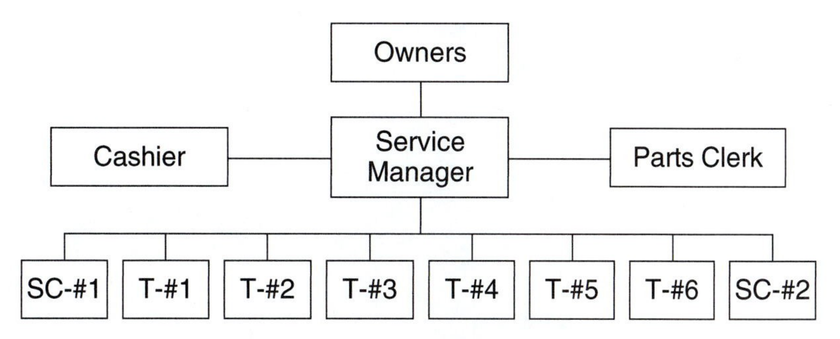

FIGURE 1-4 Organizational structure.

payable clerk (an employee who pays the invoices received by the facility). For reasons of internal control (discussed in the third section of this book), the business manager reports to the owners. In any case, the structure gives the owners less reliance on one person and greater control over the various operations by having three managers report to them as opposed to one.

Another benefit of this reorganization is that it permits the service facility to continue to grow through the addition of a new group of technicians and a service consultant. If the number of customers becomes too great for this structure or specialty services are added (such as a detailing service) or if a division is created to maintain the facility, a service director could be hired to oversee more managers. In other words, growth requires change and a structure that can be expanded, as shown in Figure 1-6, in which three of the assistant service managers' stations are seen to the right, with a cashier's area at the top of the picture.

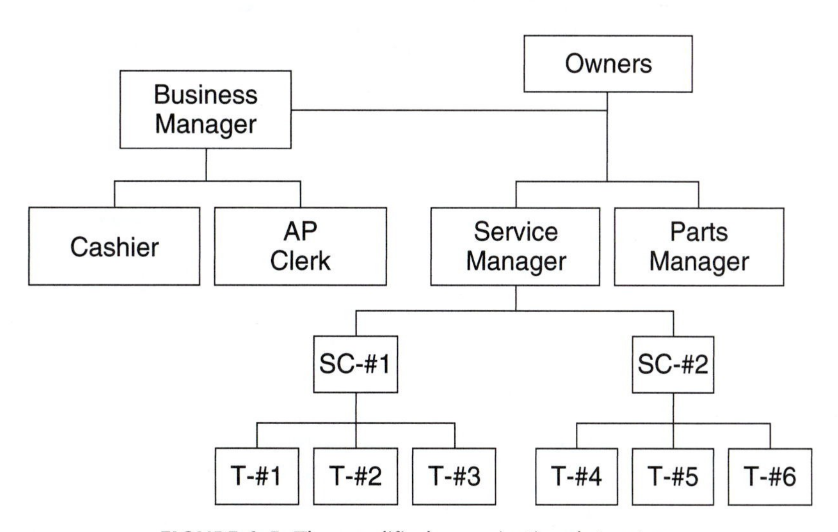

FIGURE 1-5 The modified organizational structure.

FIGURE 1-6 Assistant service manager stations.

Business Communications

An **organizational diagram** also presents the formal lines of communication as well a clear chain of command. As the arrows in Figure 1-3 indicate, when a communication goes down through the organization, the owners present information (directives, initiatives, questions, policies, procedures, ideas, etc.) to the service director. The director then relays this to the managers under his or her supervision. Next, the managers inform the employees that they supervise.

When information (via memos, notations, e-mails, or verbal messages) is sent up through the organization, it works its way up to the level of authority needed to give an answer. For example, a technician may tell an assistant service manager (service consultant) about a concern. If the assistant manager cannot provide a response, the message would then go to the service manager. The service manager may then act on the information (solve the problem or provide an answer) or send it higher if the action to be taken is outside of his or her level of authority. In some cases, the information may work its way up to the owners.

Of course, as information is sent up or down the chain of command, a manager along the way may change the message or only present part of it. The change may be seen as a way to get action or a response more quickly and, therefore, result in greater productively. Naturally, in these cases the judgment of the manager is critical because the final decision maker, such as the owners or service director, may want or need all of the information (suggestions, complaints, irrelevant details) to make a decision.

The transmission of accurate and relevant information is important to decision making. When details or problems are not included, or are inaccurate, or are received out of sequence, a wrong decision is likely. When this occurs, the owners eventually will lose confidence in the manager. Therefore, knowing the organizational chain, understanding the limits of one's authority to make decisions, and being able to properly communicate a problem and relevant facts are important to a manager's job security and career.

In some cases, communication in an organization will flow laterally as people of equal status in the chain of command must communicate with each other. In most cases, the assistant service managers (service consultants) will communicate with each other directly during daily operations, as opposed to going through the service manager. For example, if the service facility ran out of a specific air conditioning part, the service consultants would share this information with each other to ensure customers were not inconvenienced, instead of sending it up and back down an organizational chain. Figure 1-7 illustrates how communication failure resulted in a job that could not be finished on time nor put back together so the customer could drive it until the parts arrive.

In other cases, workers in one work area may communicate directly with a manager of another area. For example, if a technician notices a need for air-conditioning parts that are no longer available from the normal supplier, this could be communicated directly to the parts manager and not involve the service manager. However, if the parts manager determines that the air-conditioning parts cannot be obtained except from a supplier that is not affiliated with the service facility, then the parts manager many need to involve the service manager. The action in this case may even need to involve the owner, especially if the needed air-conditioning parts violate a pre-existing contract in a franchise agreement with the automobile manufacturer.

In most cases, determination about when and how departments should communicate with each other is the job of the managers. To help facilitate these interactions and so that each manager has the opportunity for input on important issues and understands why decisions were made,

FIGURE 1-7 An unfinished job due to lack of communication among employees.

managers should meet periodically. The relevant results of these meetings should be shared with the employees in the work units. These meetings with employees cannot be long because of the cost to the company and to those technicians who work on flat rate, but meetings are needed to coordinate the activities of the units in the facility. In addition, these meetings give the employees an opportunity to discuss specific problems that they are encountering.

An example of the benefit of a managers' meeting at a dealership regards a widespread recall for a certain make and model of automobile. In these cases, it is not unusual for sales, service, and parts managers to discuss the implications as they relate to their unit. For instance, service managers are naturally interested in ensuring enough technician time is available to take care of all of the forecasted recalls. The parts managers must make arrangements to have enough parts on hand to maximize technician efficiency and avoid downtime, which can cause a loss in profits. At the same time, sales managers need to have the automobiles in their inventory repaired immediately and inform customers who recently bought a recalled automobile of the notice they will receive. Each manager then needs to inform their employees and discuss a plan so that they know what must be done. Organizations that can effectively work together are recognized for their efforts, as seen in Figure 1-8.

Maintaining the Lines of Communication

A problem in some businesses is that the communication channels may be misused, abused, or not recognized. The line of communication must be known and understood by each employee. Furthermore, the line should be as direct as possible and involve as few managers as possible. The transmission of information naturally takes time and costs money. When too many managers are in the line, one danger is that the content of a verbal message from the beginning to the end of the line may change. Another concern is the integrity of the line, meaning the danger of information leaking from the communication channel. Because of these dangers, some facilities may want their managers to receive and transmit information personally or to ensure that the employee delegated to transmit the information can be trusted.

When a verbal message is received, the manager must carefully listen to it and properly record the information presented (do not rely on memory). Next, when transmitting information verbally, they must ensure the information is correctly presented and understood. This may require such techniques as having the person explain his/her understanding of the message or asking people in a meeting to repeat what was said (this helps to get the attention of people who are not listening). If a manager cannot transmit information verbally, he/she must select people that can be trusted to do it properly.

FIGURE 1-8 Awards presented to an effective organization.

Finally, some information being transmitted may be confidential, such as information about an employee's performance or union negotiations. Even if the information is not confidential, it should only be made public by the manager who the owners authorized to make the announcement. As a result, in many cases a manager needs to limit the number of people who have access to the information or documents. Documents transmitted through channels should not lie on a desk for others to read.

An example of the need to follow proper communication channels involves a service manager who was told that the manufacturer had stopped reimbursing the dealership for rental cars given to customers. The service manager told the dealership controller (accountant) to look into it. The controller confirmed that the dealership was not reimbursed for rental cars that were not under warranty or whose repair took longer than five days.

In response to this information, the controller sent a memo directly (not through the service manager) to each service consultant. He explained that he would take the cost of a rental car out of the service consultant's paycheck if the dealership was not reimbursed by the manufacturer. Because the service consultants did not understand the parameters of the rental car policy, they did not give any rental cars to any customers for fear of making a mistake. The customers became unhappy and complained to the manufacturer, some even threatened to sue the manufacturer. The manufacturer then threatened to sue the dealership. Then the dealership owners threatened to fire the service manager.

In this case, there were several mistakes. First, the service manager should have included the owner in the meeting with the controller. Second, the memo by the controller should have been approved by the owner, who then should have passed it on to the service manager to present to the service consultants. Finally, as a staff manager, the controller exceeded his authority and should have known better. In any event, the way in which the problem was handled caused harm to the reputation of the service manager and the service facility.

Concluding Thoughts

Too often, service managers are deemed unsuccessful or ineffective because they do not respect some basic business rules, principles, or protocols (meaning code of behaviors, recognition of rank or precedence, and the following of resolutions). Such failures may not be an accurate interpretation of the manager's knowledge and ability but may be a result of not being aware of expectations, not knowing how to follow proper procedures, or realize that recognition and practice are important. A person may be an excellent service manager but may not fit into the business setting of the facility as a result of limited understanding. This, of course, can be corrected through the careful study of the facility's history and practices and the job expectations.

In addition, owners and professional service managers know that the use of the chain of command and use of communication channels are critical when service facilities are busy and the pressure is on everyone. When the formal lines of authority and communication are not followed as a regular practice, however, employees will not be able to use them as intended when the pressure is on. Confusion and errors will occur. This often leads to a change in managers.

Of course, there will always be those employees who will not follow the chain of command or lines of communication. This also causes confusion and problems. Too often disorder and errors lose customers, reduce profits, and result in employee turnover. Regardless of how well employees are liked or how good they are at doing their job, the service manager's only option is to terminate their employment if they cannot or will not follow the rules of operations. These employees are too risky to retain and give a wrong message to those employees who follow the practices and procedures established by the owners of the business.

Review Questions

1. Describe how to prepare a diagram for a management structure at an automotive service facility and draw an example for a typical dealership.
2. Explain the difference between line and staff positions and give examples of each at a typical service facility or dealership.
3. Describe how communication flows at a service facility and in what situations might it change and why.
4. Explain what would be meant by job authority and provide examples.
5. Describe how an organizational structure is related to the chain of command.
6. Draw an example of a flat span of control and explain in what situations it can be problematic.

CHAPTER 2

OWNERSHIP, TYPES OF FACILITIES, AND THE SERVICE MANAGER'S JOB

LEARNING OBJECTIVES

Upon reading this chapter, students should be able to:

- *Describe the basic features of the following types of ownership:*
  - *proprietorship*
  - *partnership*
  - *corporate ownership*
  - *limited liability company*
- *Explain the purpose of the income statement and balance sheet.*
- *Calculate a company's profit given its income and expenses.*
- *Explain what asset, liability, and owner's equity mean.*
- *Calculate an owner's equity given the amount of the company's assets and liabilities.*
- *Describe the purpose of the six basic types of automobile service facilities.*
- *Prepare a flow diagram for a basic management structure for each type of automotive service facility.*

Introduction

Successful service managers need to know who owns their facility and how it is owned (proprietorship, partnership, corporation, limited liability company). The differences among the types of ownerships must be recognized because of the way owners invest their money in the business. In addition, the means of ownership affects the decision making process. For example, an owner of service facility may be the sole investor of the business (owns 100 percent) or may only own a small percentage (owns 10 percent) of the facility. Furthermore, the type of ownership may place the personal assets of the owner, such as their residence, at risk as well as their investment, while the risk in other types of ownerships is limited to the money investment. Such a difference, of course, could influence the interactions between an owner and employees and also the owner's involvement in the management of the facility.

Recognizing the different type of service facilities is also important to service managers. For instance, the type of services sold by a facility will likely influence the **organizational structure**, the chain of command, and the authority assigned to a service manager. To discuss these different types of automobile service facilities, the chapter divides them into six categories.

Therefore, there are a variety of service facilities because of the different methods of ownership and different types of facilities. Although some combinations may be the same, they still may not operate alike because of their different management structures, policies, rules, procedures, and protocols. The one feature that all service facilities have in common, however, is the way in which they calculate their profits and the worth of the business. As a result, this chapter introduces basic financial terms and statements, which are further discussed in the third part of the book. An understanding of the use these terms and statements is critical to the financial responsibilities of service managers.

Basic Forms of Ownership

There are several ways that a person may own or have ownership in a service facility. Each has legal advantages and disadvantages; however, the legal details are beyond the scope of this book. Of importance, however, are the legal and business influences of the method of ownership on the personnel and financial operations of a facility. The differences among them are significant to the managers and employees of a business. The kinds of ownership presented here include proprietorships, partnerships, corporations, and limited liability companies.

Proprietorship

An automotive service facility owned by one person is called a **proprietorship**. The proprietor is the sole owner, who provides the money needed to open, buy the equipment, and maintain initial operations of the business. A proprietor receives all of the profit earned by the facility and is personally liable for the facility's *debts*.

Proprietors can operate a business under their name (e.g., John Smith's Auto Service) and can use a personal checking account to deposit money and pay business debts. Proprietors may also register for a fictitious name in their state and then run it under that name, such as *The Car Doctor*. In this case, John Smith would have to file the proper paperwork to ensure he has not infringed on a protected trademark or an operating corporate name. Then his checking account would be opened as The Car Doctor. John Smith could receive checks made out to The Car Doctor; however, to deposit them, he would have to endorse them by signing the back of the checks as follows: John Smith DBA (doing business as) or TA (trading as) The Car Doctor. When writing a check for the business, Smith would have to sign his name. In other legal documents, such as tax reports, Smith also would have to sign his name and then either DBA or TA followed by the name of the business.

As a proprietorship, if the bills of a facility cannot be paid, the owner is personally responsible. In other words, his/her personal property could be seized (in some states, if the personal property is jointly owned with someone who does not have an ownership in the same business, then it cannot be seized). The personal property can then be sold to pay any *creditors* to whom the facility owes money.

When a profit is earned by a facility under a proprietorship, all of it must be claimed by the owner at the end of the 12-month business year (known as the **fiscal year**). Company profit is earned when the company's income is greater than its expenses. The owner receives this profit as the payment in return for his/her work and on the money invested in the business. The profit (or loss) for a facility is reported on the owner's personal income tax return. This means that the proprietor's business does not pay a tax on the profits earned. The profit or loss of a company, regardless of ownership, is presented on a financial statement known as the **income statement,** which presents the amount of money earned minus the expenses to show the profit or loss for a specified period of time, such as a week, month, or year. The basic formula for the income statement is:

Income − Expenses = Profit

The income statement is described further in the last section of this book.

Some proprietors, or owners, may draw a regular salary (particularly for larger proprietorships when taxes must be paid to governments quarterly). These drawings are not a company expense that reduces the company profit but, rather, an amount that reduces the profit that will be received by the owner at the end of the year. As a result, the amount of money taken by the owner on a monthly basis and the amount taken at the end of the business year equals the company's profit. For example, assume an owner takes a salary or drawing of $4,000 a month, or $48,000 a year. The owner must pay income taxes and the employee's (the owner) and employer's share of Social Security/Medicare taxes (known as **FICA** or the **Federal Insurance Contribution Act**) on the $4,000 each month. Consequently, if the company has a profit of $100,000 at the end of the year, the owner would receive an additional $52,000 ($100,000 − $48,000 of drawing = $52,000). The owner would have pay income taxes on the $52,000 which includes the employers and the employee's share of the Social Security/Medicare taxes. So, the proprietor will have to pay the income taxes on $100,000. Of course, if the business loses money, the owner would claim the loss.

In some cases, a proprietor may not wish to withdraw all of the profits in order to expand the business or purchase new equipment. In these cases, the proprietor would make the purchase during the year and the amount would reduce the profit of the company. For example, assume a business decided to purchase a piece of equipment for $10,000. An end-of-year profit of $50,000 would then be reduced by $10,000 to $40,000 ($50,000 − $10,000 = $40,000). This deduction assumes the tax code at the time allows for a 100 percent write-off of the capital purchase. If not, the amount paid for the equipment would have to be **depreciated** and a smaller amount, such as $2,000, would be charged against the profit each year (calculations for depreciation are explained in the last section of this book).

Note that when a business pays personal salaries and sends the taxes withheld and FICA payments to the federal and state governments, it uses an **Employer Identification Number,** known as an **EIN**. An EIN is received from the federal government and needed by service managers for other documents.

If a proprietor would like to set aside any of the company profit into an emergency or savings account, it must be set aside by the owner *after* the owner has paid the taxes on the money. In other words, the profit earned by a company cannot be placed in a savings account or invested by the business. This can be done by the owner but only after taxes, including FICA taxes, have been paid (retirement plans excluded).

When a proprietor closes a service facility and sells the equipment and property, the proprietor receives the balance of the money left over after the debts of the facility are paid. The difference between the amount of money an owner receives for the sale of a business less the company debt is the profit on the sale, which is referred to as a capital gain. The proprietor must then pay a capital gains tax on the profit from the sale of the business. This gain, as shown in Figure 2-1, is

Amount received from the sale of a business	$500,000
Less company debt	350,000
Profit on the sale	$150,000

FIGURE 2-1 Sale of business.

calculated from the amount of the profit from the sale and is what the company is worth.

Therefore, calculating what a company owns, known as **assets,** and subtracting what the company owes, known as **liabilities**, gives the amount of the company's worth, known as the **owner's equity**. A **balance sheet** is a financial statement that presents a list of a company's assets (what is owned), a list of the company's liabilities (what is owed), and the owner's equity (what the company is worth). The accounting formula used to prepare the balance sheet, which shows a company's assets equals a company's liabilities plus the owner's equity, is:

Assets = Liabilities + Owner's Equity.

Because the focus or point of discussion is often on the owner's equity (as shown in Figure 2-1), the formula is often discussed in terms of company assets less liabilities equal owner's equity or:

Assets − Liabilities = Owner's Equity.

The balance sheet, assets, liabilities, and owner's equity are discussed further in the third part of this book.

The financial statements (balance sheet and income statement) of a business are important documents to all companies, regardless of ownership. For example, before a bank grants a loan to a business, the financial statements will be studied to determine whether a business has sufficient equity and is making a profit. A bank does not want to risk losing any money if a business must be sold or **liquidated** (meaning everything owned by the business must be sold to pay all debts). Therefore, the amount of a loan given by a bank depends on the amount of the owner's equity and profitability.

When profits are reinvested by purchasing equipment for a business (increasing assets) or reducing the debt (reducing liabilities) there is an increase in the owner's equity. When a loss occurs and more money is spent than earned, the cash account is reduced (decreasing assets) and bills cannot be paid or loans are made (increasing liabilities), the owner's equity is reduced. When cash is not available to pay the debts of the business, the owner or owners may have to put more money into the business. In other words, the owner's equity and owner's profit are directly impacted by a company's profit or loss. This is a problem for the managers because all owners want to be paid for their work, receive a return on the money invested in the company, and show company growth via an increase in owner's equity. Consequently, service managers must constantly monitor sales (revenue or income) and the cost of sales (cost of labor and parts to fix a car) to ensure a gross profit is earned on each job. The gross profit must be large enough so that the other expenses (salaries of manager, rent, heat, light, etc.) can be paid and a net profit can be earned. Such an assignment may seem simple, but it can be quite complex.

Partnership

A **partnership** operates like the proprietorship described earlier except that two or more people own the facility. Like a proprietor, the partners receive the profit earned (or claim the loss) by the service facility. Like a proprietorship, the profit is not taxed but is distributed to the owners who report it (or a loss) on their personal tax returns. In addition, the partners are personally liable for the debts of the facility and, like a proprietorship, their personal property could be taken to pay them. Limited partnerships and limited liability partnerships are similar to limited liability corporations, discussed later, in which the owners may not be held liable for the debts and other liabilities of a company.

In most cases (unless contractually agreed upon before business startup), the amount a partner invests in a company determines the percentage of profits each partner can claim. For example, if a partner put up 60 percent of the money for a business, that partner would receive 60 percent of the profits or have to cover 60 percent of a loss. In another example of a partnership, assume that a business is owned by four investors. Partner A invested $50,000, partner B invested $50,000, Partner C invested $50,000, and Partner D invested $25,000. In this example, each partner was given an equal share (25 percent) of the business, including Partner D, who was a master technician. If the business makes a profit of $40,000, each partner would receive $10,000.

Corporation

A **corporation** is considered a legal entity. This means it is treated like a person (entity) and, therefore, may own property, must pay taxes (at the corporate tax rate), can incur debt, can sue a person or another corporation, and can be sued or held liable for crimes. The money needed to start a corporation comes from the sale of *shares of stock* issued by the company to investors called **stockholders**. The number of shares issued depends on how much money is needed and how much is charged in the sale of a share of stock. As a result, the cost of a share of stock varies from corporation to corporation.

Unlike proprietors or general partners, stockholders are not personally liable for the corporation's debts or actions that may be illegal (however, the corporation's officers—president, vice president, treasurer, and secretary—can be held liable). Investors who buy stock, however, expect the managers to use their money effectively and efficiently to earn a profit. If not, they can sell their stock and invest it in another company. When shares of stock are sold, the amount paid is determined by an agreement between the seller and buyer.

The percentage of ownership in a corporation is determined by the number of shares owned divided by the number of shares the corporation issued. As the company grows and generates profits, the value of the shares should increase. Of course, the value of the stock may decline when a company has a loss or encounters unfavorable business conditions. For example, if one of a corporation's service facilities has to close because it was not showing a profit, the value of the stock will likely decrease.

When a corporation is created, a meeting of the owners is held to ratify (this means to approve) the articles of incorporation formally. They also issue shares of stock to all investors and possibly for people to buy, elect the officers of the corporation, approve a set of bylaws, and authorize the opening of a bank account. The shareholders must then meet on an annual basis. The rules and regulations that govern the corporation are the **corporate bylaws**. These bylaws usually include provisions regarding the election by the shareholders of the **corporate directors**, who are often referred to as the Board of Directors or Corporate Board. The corporate bylaws also address the rights and obligations of shareholders, appointment of corporate officers, and general corporate business matters. The bylaws lead to the creation of corporate policies that, in turn, eventually lead to the preparation of the operational procedures and regulations that are important to the service manager, as discussed in Chapter 4.

The directors of the corporation make major corporate business decisions, but likely the most important task is the appointment of the **corporate officers** and the monitoring of their performances. The corporate officers include a president, vice president, treasurer, and secretary. In some cases, one person may hold more than one position. The officers are then responsible for the oversight and management of the daily operations of the corporation.

Unlike a proprietorship or partnership, corporations must pay taxes on their profit. After the taxes are calculated, some of the profit (also referred to as earnings) may be distributed to the shareholders. This distribution is referred to as a **dividend**. The stockholders must then pay taxes on their dividends. This is the reason that some people believe that corporate profits are double taxed (once on the profit and once on the dividends earned from the profit). Changes in the tax laws have been sought by investors for a number of years.

For example, if a corporation has earnings (profit) of $4 per share, it must pay tax on the total earnings ($4 times the number of shares). At the same time, if the board declared that $1 of the $4 would be paid as a dividend, a person who owned 100 shares would receive a dividend payment of $100. The shareholders must then pay taxes on the dividend payment received. The remaining $3 of the earnings (profit) would be retained by the corporation (called *retained earnings*) and could be used to expand cash reserves or for reinvestment in the business for expansion or the early repayment of a debt. The board determines the amount of earnings retained by the corporation. A corporate board, however, may not pay any dividends to the shareholders but would expect investors to reap advantages from their investment by an increase in the value of their shares. Regardless of whether the corporation pays or does not pay a dividend, it must pay taxes on all earnings (profit) even when the money is reinvested in the corporation.

S Corporations

The corporation described in the previous section is referred to as a **C Corporation**, and the earnings of these corporations are taxed under subsection C of the Internal Revenue Service (IRS) code. An **S Corporation** however, comes under subchapter S, which allows the corporation to avoid paying taxes on earnings. In other words, an S Corporation, referred to as an S Corp, is treated like a proprietorship and partnership, in which all of the business profits must be distributed to the shareholders at the end of the year. Then, the owners must pay personal income taxes on the dividends received from the corporation. Unlike a proprietorship or partnership, however, the shareholders of an S Corp are not personally liable for the debt of the facility because it is considered a legal entity.

Limited Liability Company (LLC)

A **Limited Liability Company (LLC)** is also considered a legal entity like a corporation. Owners of an LLC are referred to as members, which may be a person or another entity like a corporation. The members of the LLC make business decisions for the company unless they specify otherwise in their articles of organization. For example, the members may determine managers will run the company. If stated in the articles of organization, the managers of the LLC may serve and have the authority of a corporate board of directors.

Like a partner or shareholder, the percentage invested by a member of an LLC determines the amount of ownership. The two major benefits of an LLC are that the members are not personally liable for the debts of the company and the business is not taxed. Rather, the profit or loss of the company is passed through to members and reported on their personal income tax returns.

Types of Automobile Service Facilities

In addition to the different methods of ownership, a manager must understand the different types of automobile service facilities. As a broad definition, an **automobile service facility** is a business that sells maintenance, repair, and diagnostic services to automobile owners. Not all automobile service facilities are the same and a service manager must respect owner expectations based on the services offered by the facility that employs them. To explain the various types of automobile service facilities, they have been classified into six categories:

- General Independent Service Facility
- Dealership Service Facility
- System Specific Service Facility
- Product Specific Service Facility
- Chain Service Facility
- Fleet Service Facility

The General Independent Service Facility

When the public thinks of an automobile service facility, the one that most often comes to mind is the general independent service facility that performs maintenance, repair, and diagnostic work on all models and makes of automobiles. These facilities range from a one-bay shop, to a service facility located in a gasoline station, to a facility with multiple bays and a large parking lot. Some independent facilities may not perform all of the services; for example, some may not rebuild or replace automobile engines. However, in general, these facilities work on all makes and models and most systems, as shown in Figure 2-2.

Because the independent facilities are not the same size and some may limit their services, the duties of a service manager will vary from facility to facility. For instance, the size of the facility dictates the number of customers and the volume of work processed on a daily basis. In turn, the amount of work processed on a daily basis determines the number of employees needed and the amount of money handled. Although the management process and systems at all independent service facilities are similar, the volume of work and complexities of the work assignments dictate different circumstances.

The Dealership Service Facility

The maintenance, repair, and diagnostic services performed by new-automobile dealerships often concentrate on specific makes and models of new and used automobiles. As a result, they are considered a **dealership service facility**. The technicians tend to specialize in performing services on the automobiles sold by the dealership, as shown at the Ford dealership service department in Figure 2-3.

Management of personnel and financial operations at a dealership service facility are similar to those at an independent facility. A more complex work environment may exist, however, because of the need to have multiple service consultants with technician work groups to be effective. In addition, because a dealership service facility handles new-car warranties, the record-keeping and financial duties of a dealership service manager are likely to be different from those of a manager at an independent facility.

FIGURE 2-2 Independent automobile service facility in Hawaii.

FIGURE 2-3 Dealership service department buys.

The goal of the service manager is still the same, which is to make the service facility profitable. Likewise, the personnel responsibilities are the same.

The System Specific Service Facility

Service facilities that have grown in popularity over the past 50 years are those that repair and maintain one automobile system. Because of their limited services, these facilities are referred to as a **system specific service facility**. For example, a system specific facility may limit its services to brakes, mufflers, transmissions, engine tune-ups, or oil changes. More recently, some facilities have focused on specialized installations and repairs, such as custom stereo systems or high-performance parts replacement.

Because of their specializations, a system specific facility is likely to have unique equipment, tools, and procedures, and even a unique shop layout. Furthermore, the skills of the technicians may range from a highly specialized system specific technician (such as a transmission specialist) to a low skill level technician (such as a technician who drains engine oil). Although these differences are obvious, the manager's personnel and financial oversight duties will be similar, although the focus will be on the area of specialization.

FIGURE 2-4 Franchised system specific service facility.

Because of the repetitive nature of a system specific service facility, which permits a high volume of work to be processed, the use of specialized technicians who may be trained at the facility, and the need to handle a limited inventory, this type of facility has become a popular business to franchise (discussed later). For example, franchised service facilities that sell system specific services include Midas, Aamco Transmission, Maaco, Mienke Muffler, Jiffy Lube, Cottman Transmission, and All Tune and Lube. Additional information is presented later in the section on franchised businesses.

Product Specific Service Facility

A service facility that diagnoses, repairs, and performs maintenance on specific makes and models of automobiles (such as Rolls Royce or Bentley) is known as a **product specific service facility**. Because their services are limited to specific automobiles, they may be similar to a new-car dealership; however, they often do not sell them (if they do, the volume is usually relatively low). Although these types of facilities are not large in number, they are popular in some regions of the country.

Because of their unique capabilities to repair a specific make of automobile, some of the expectations of a service manager will likely be different. For example, a manager needs to have an extensive knowledge of the particular type of automobiles serviced and the types of customers who own them. The manager also must hire, and often arrange to train, technicians who can work on automobiles with out-of-the-ordinary systems. For example, the classic Rolls Royce Silver Cloud has a brake system

FIGURE 2-5 Product specific luxury import service facility.

with a mechanical clutch (servo) driven off the tail shaft of the transmission that uses a complex arrangement of rods to actuate the master cylinders. The adjustments, repairs, and overhaul procedures needed to ensure the system both works as designed and will last, plus the workmanship and care expected by the owners of these Rolls Royce automobiles, can only be obtained on the job from a Rolls Royce Master Craftsman. As a result, a manager of a Rolls Royce mechanical restoration shop must have the knowledge of the specialty automobile to convince an owner that the automobile can and will be repaired and that the charges for the service are justified (a Silver Cloud complete brake system overhaul with replacement parts can cost up to a few thousand dollars). Therefore, although the basic personnel and financial responsibilities of service managers at product specific facilities are similar to those at other facilities, they have special challenges with respect to technician recruitment, training, and customer service.

Chain Service Facility

A service facility that is one of several facilities owned by a business is called a **chain service facility** (such as Sears, Roebuck & Co; Wal-Mart; and Pep Boys). The chain service facility should not be confused with a franchise service facility owned by independent local owners who pay for the rights to use a name and sell specialty services from a corporate franchise dealer. The chain service facility is different because it is not owned locally. In addition, the facility operates within a larger corporate organizational structure; each facility sells the same services and parts, and follows the same procedures. In some cases, a chain service facility may offer

FIGURE 2-6 Chain service facility.

a wide variety of services, while another may provide limited maintenance work and the sale and installation of specific products, such as tires and batteries.

Because the manager of a chain service facility is part of a larger organizational structure, the position is similar to one at a dealership service facility. For example, the financial and human resources responsibilities of the service facility may offer more assistance, but they also must fit into the operations of the dealership or a national corporate office. In most cases, the service manager of a chain service facility reports to the store manager, or a regional service manager, who in turn report to a manager in a national or regional corporate office.

Fleet Service Facility

A facility limited to servicing automobiles owned by one company or government agency is called a **fleet service facility**. The technicians at a fleet service facility may provide all of the services offered by a general independent service facility or be limited to maintenance work. The reasons to limit service work may include any of the following:

- The fleet has a policy of trading vehicles within a given time frame or after reaching a number of vehicle miles, such as 3 years or 60,000 miles, whichever comes first.
- Most fleet automobiles are newer models covered by manufacturer warranties and/or extended warranties so that repairs are performed at the manufacturer's dealership.
- The fleet service facility employs personnel who are limited to maintenance work and basic repairs while an independent service facility per-

forms all major repairs that require technicians with advanced training and the use of expensive equipment or special tools.

Fleet service facilities are divided into two types, which are based on whether the owners are nonprofit or government agencies or by profit companies. Nonprofit and government organizations include hospitals, churches, transportation services for senior citizens, fire departments, police departments, school district buses and vans, and so on. Usually the services provided by the employees of the fleet service facility are limited. When a repair is not under a warranty or within the ability of the technicians employed, the job is sent to a specialty service facility (such as transmission repairs under a sublet repair contract), dealership, or general independent service facility.

The most important difference between a service manager's position in a nonprofit or government fleet service facility and all other managers is that the manager is not expected to generate a profit. Rather, when a nonprofit company or government agency owns the fleet, the service manager must keep the facility's cost within the amount of money allocated for its operation, which is referred to as a **budget**. If a manager cannot stay within the budget allocation, the cost to operate a fleet facility may outweigh the convenience and benefits. In these cases, the fleet service facility would be closed and all services would be contracted out to an independent service facilities or dealership service department.

When a for-profit company (such as UPS) operates a fleet service facility, then the service facility is expected to provide convenience and keep the cost of the facility at an amount where profits are increased. The manager of the facility may also work within a budgeted amount; however, as

FIGURE 2-7 Government fleet service facility.

FIGURE 2-8 For-Profit fleet service facility.

the business operations increase, the amount spent by the facility also will increase. A for-profit fleet manager may have to keep the costs per vehicle or cost per mile within a set of limits that will benefit company profits. This requires the service manager to be concerned about company profits in a slightly different manner than managers of other service facilities.

Controlling the costs of a fleet service facility for a profit company may be accomplished in a number of ways. Several of these are to monitor the costs of the services to each fleet vehicle (labor and parts), reduce costly vehicle downtime during the service process, and implement a program that will extend the life of each vehicle. These all require the fleet service manager to collect and use the data that will track the costs to create and implement cost controls and to compare them to the same costs at an independent service facility. The fleet service manager also must keep track of repairs and repair times to ensure that each vehicle is ready to go by the start of a new shift (reduce downtime), help avoid costly breakdowns (keep the vehicles on the road), and extend the lives of the vehicles (reduce the cost of purchasing new vehicles). Data collection and analysis are important skills and, as such, make a fleet service manager's job different from those found at other types of service facilities.

System Specific Franchise

As mentioned earlier, a system specific service facility such as AAMCO, which works on transmissions, or Jiffy Lube, which performs oil changes, is a franchise facility. These facilities are operated by a local owner or owners who purchase the rights and then pay a monthly fee to a national franchise in return for the use of their name. The national franchise, meaning the one that sells the use of the name, is usually a C Corporation, and the local owner or owners are under personal contract to the national franchise. Consequently, the national franchise office must approve the local franchise owner or owners. If approved, the local franchised business then usually operates as a C or S Corporation or LLC.

In exchange for the initial fee and contract to pay a monthly fee typically as a percentage of the monthly gross sales, the local franchise owner or owners are provided with a protected geographic area. This means the national franchise will not locate another local franchise within the region. The home or regional franchise office usually has to approve the building and location of the service facility in the geographic area. After the local owners and location are approved and the business is ready to be opened, the local owners and managers may have to attend training sessions at a national or regional office.

Representatives of the franchise typically observe the opening of the local business and then make regular visits (some unannounced) to inspect the facility. For example, a franchise may require a local business to follow a script when answering the phone and greeting customers. Announced visits and cold calls (in which a representative pretending to be a customer calls on the phone or comes into the local business) may be made by the national office to check out their use of the scripts and customer treatment.

A franchise also may provide local owners with a variety of services and assistance, such as a business system, business consulting services, agreements to supply parts and equipment, advertising and marketing assistance (often mandatory and for a fee), as well as ongoing training. Some franchises also may require their local businesses to use standardized forms to record and report all sales and expenses. One reason may be because the fee paid to the national franchise is a percentage of local gross sales. The use of standardized forms attempt to insure that local businesses report all sales. Another reason is that the forms permit the national franchise office to track all sales of all local businesses. For instance, if the sales of a local business has declined over a period of time, the home office may send investigators into the area. Possibly the investigators will find that the local owners are not keeping up their end of the contract and running the business as per the expectations of the national or regional managers. In such cases, the national office has a contractual ability to close the local franchise.

Product Franchise

Essentially, an automobile dealership is a franchise that operates under a slightly different agreement from the service franchise described earlier. For this reason, they are considered a product franchise because they are based on agreements to sell a manufacturer's products, specifically automobiles and parts. The basis for the contract is not an agreement to provide a percentage of gross sales to the manufacturer but, rather, for the manufacturer to generate sales income when the dealer sells one of its products (automobiles and parts).

FIGURE 2-9 Product franchise.

The service department of a dealership operates under the product franchise agreement and, as such, presents a number of provisions for the service manager to follow. For instance, the service department of the dealership must perform all warranty work required by one of their automobiles, regardless of the dealership from whom it was purchased. The agreements also specify that the service department use, and the parts department sell, only parts supplied by the manufacturer. Furthermore, the service department must meet the standards set forth by the manufacturer for technician training, special service tools and equipment, and customer service as measured by scores for FFV (fixed first visit) and CSI (customer satisfaction index), as well as other assessments. Finally, dealership product franchise agreements set forth a number of expectations for service managers to follow, including operational procedures, the submission of paperwork, financial reports, and so on. The service manager also must work with the manufacturer's representatives when necessary to improve services and to take care of any customer concerns about the products purchased.

Illustration of Organizational Structures

In theory, there may be as many organizational diagrams as there are service facilities. No one diagram fits all service facilities but, rather, they must be designed to fit each one. To make this point, two very different organizational diagrams are presented here.

The first diagram (Figure 2-10) presents a more elaborate structure for a dealership. In the figure, a general manager, a controller, and a human resources manager report to a president, who reports to a corporate board. The diagram then shows four line management positions under the general manager (positions under the controller and human resources are not included in the diagram). In such a dealership, the title for the general manager could be vice president because the person is basically responsible for all line operations (sales, service, and parts plus the receipt of money into the business). The controller is responsible for the payment of the bills and payroll, while personnel comes under a separate manager.

At the same time, in this dealership (Figure 2-10), the service manager oversees the service operations of the facility. This requires the service manager to supervise the service consultants (who also may be referred to as assistant service managers) who, in turn, supervise the technicians (some technicians may be staff managers and referred to as lead technicians because they assign work to the technicians on their team). In the diagram, the service manager, sales manager, business manager, and parts manager are given equal status. In other words, the diagram in Figure 2-10 presents a relatively large corporate enterprise with a number of managers and specialists who run the business.

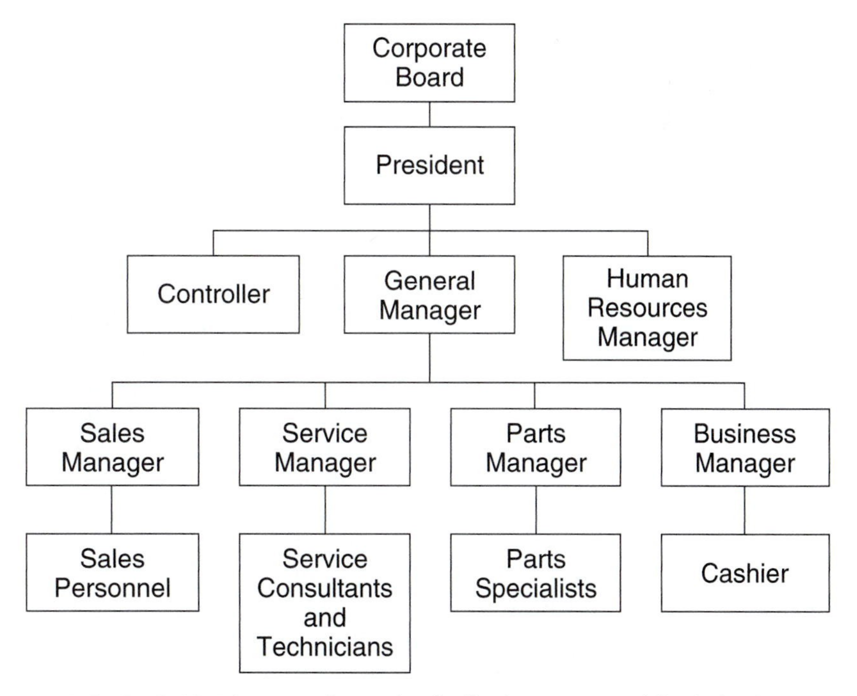

FIGURE 2-10 Diagram of a service facility in an automobile dealership.

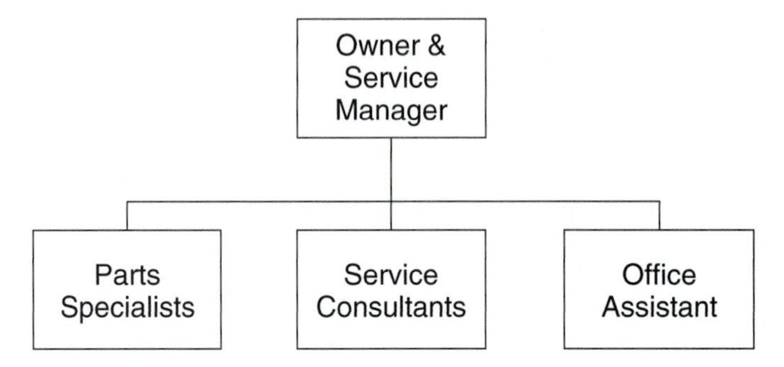

FIGURE 2-11 Service facility diagram with owner/service manager.

At the other extreme, Figure 2-11 shows an independent repair facility operated under a proprietorship. In this diagram, the owner serves as the service manager and oversees a parts specialist, service consultant, and office assistant. The owner is directly responsible for overseeing people who have a variety of duties from waiting on the customers, working with technicians, and purchasing parts, to collecting money. In this arrangement, the owner is likely to take care of the finances (write checks, make deposits, track debt and credit), personnel (hiring to appraisals), advertising, tax payments, and so on. Although this arrangement is quite common, there are lot more organizational diagrams that exist for facilities between the smaller one shown in Figure 2-11 and the dealership in Figure 2-10. One diagram cannot fit all types of service facilities.

Concluding Thoughts

The first two chapters have shown why service managers serve as the pivot in a service facility. Basically, they stand between upper management and the people who conduct the actual day-to-day work repairing automobiles. Service managers are the ones who ultimately deal with shop personnel, financial activities at the customer level, as well as the numerous salespeople and vendors who come to a facility unannounced. The service manager's job is varied, busy, and requires a lot of people skills.

In order to fulfill the many duties and expectations of such a position, the service manager must have a firm grasp of the basic features of a facility. Knowing who owns the facility, how it is owned, who the boss or bosses are, the type of work to be conducted by the people in the facility, and how to properly and effectively communicate within the management structure is critical to the assumption of the authority and the understanding of the responsibility assigned to the service manager.

These must be recognized by the service manager when taking a job. Making incorrect assumptions can be a deadly mistake.

Next, in order to be an effective manager within an organization, managers must study the facility's business plan. The business plan is the game plan and the topic of the next chapter. Too often, the plan is not known, understood, or used by a service manager. Like any game plan, a person cannot provide leadership for a team if he or she does not know the plan and what he or she is to do as a leader of the team. A team leader must know who is the coach, who is in charge (the owner), who calls the plays (general manager) and how they are called, and what team members are supposed to do (service consultants and technicians).

Review Questions

1. Describe the basic features of the following types of service facility ownership:
   A. proprietorship
   B. partnership
   C. corporate ownership
   D. limited liability company
2. Explain why managers and owners use an income statement and balance sheet.
3. Describe the relationship among income, profit, and expenses and show how these mathematical relationships are used to calculate profit.
4. Explain the difference between an asset, liability and owner's equity and give examples of each.
5. Illustrate the mathematical calculation made to compute the amount of an owner's equity.
6. Describe how each of the six basic types of automobile service facilities differ and give examples of each in your town or city.
7. Draw a diagram that shows the basic management structure for each type of automotive service facility.

CHAPTER 3

STRATEGIC BUSINESS PLANNING

LEARNING OBJECTIVES

Upon reading this chapter, students should be able to:

- *Outline a business plan.*
- *Describe the purpose of strategic, tactical, and operational plans.*
- *Prepare mission, goal, and objective statements for a service facility.*
- *Prepare operational goals and objectives for a service unit in a facility.*
- *Explain why a management budget is important to tactical and operational plans.*
- *Explain how and why the environment affects a business plan.*

Introduction

After a service manager understands a facility's ownership and type of business (Chapter 2), and the organizational structure and chain of command (Chapter 1), he/she is now ready to study and work with a **strategic business plan**. Regardless of method of ownership, type of service offered, or size, each facility must have a *business plan*; however, most business plans are not a strategic plan. As this chapter will explain, this is unfortunate, since the business plan should begin with a **strategic plan**. This is the reason it is called a strategic business plan. The service manager, who is a team leader, must be well versed in as many details of this plan as possible.

After the entire strategic business plan has been completed, the company prepares their operational procedures and operation manuals (described in the next chapter) for the facility or for each division, department, office, and work unit in the facility. In other words, as shown in Figure 3-1, an operation manual (rules and regulations on how work will be conducted or processed) for each unit in a facility depends upon the operational procedures (how work flows through the units and officials in a facility). Operational procedures, in turn, depend on the strategic business plan (which describes the type of ownership, purpose of the business, what is to be achieved, organizational chart, goals, objectives, plus other information about the company). In other words, this alignment ties the work that will be conducted in a unit to the reason the company was established.

The preparation of the plans, procedures, and manuals takes time; however, running a company cannot involve guesswork. Everyone must know how the company wants it done and the carefully prepared rules

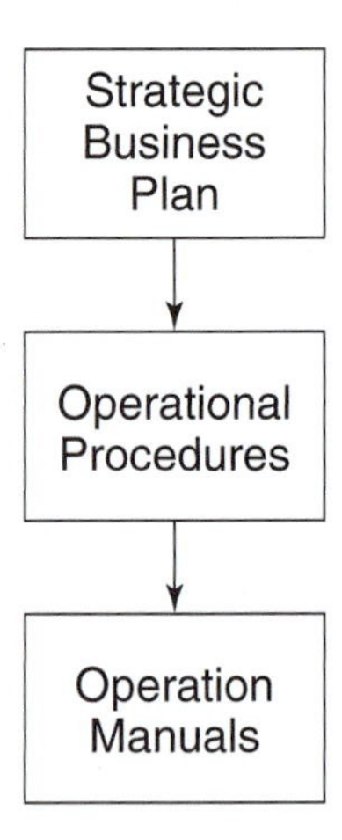

FIGURE 3-1 Connecting the business plan to unit operations.

and procedures help guide employee actions and decisions. Managers also use the rules and procedures to meet their responsibilities and lead their workers in a manner consistent with the strategic plan. Because managers do not oversee every employee action, workers must be able to refer to an operation manual when they have questions about what and how they are to do something.

The strategic business plan is required to:

- Attract investors
- Give to banks with loan applications
- Include in requests for government financial support
- Show to other businesses with applications for credit
- Discuss in the recruitment of top managers
- Help guide the company decision makers.

Therefore, the plan should not be long or complex. Rather, it must be clear, to the point, organized, and easy to comprehend. It can be as short as a page (for a small proprietorship) or contain several pages (for a corporation). It should be optimistic and forward-looking. After it is prepared, it should be reviewed thoroughly by all managers. At the end of each fiscal year, the plan should be reviewed and, if necessary, updated. The purpose of this chapter is to present an outline for a strategic business plan and to explore how environmental factors can affect it.

Business Plan Outline

There are numerous books about how to prepare a business plan and what should be in it. Regardless of the format or content, the major purpose of a business plan is to make the owners and managers think ahead and make decisions before stepping onto the floor of the business.

A suggested outline for a strategic business plan is shown in Figure 3-2. In this outline, the first item in the first step calls for the preparation of a strategic plan of the business. Preparation of the strategic plan and then the subplans (tactical and operational) is often avoided because owners and managers are required to make tough decisions, answer difficult questions, and make commitments.

The information presented in a business plan cannot be provided after the business opens. This is too late. Managers and workers must know what they are expected to do and accomplish before they start their jobs. In addition, many, if not all, of the statements made in a plan become the basic policies of the company. As discussed in the next chapter, company policies lead to the directives presented in an operation manual. Therefore, they require deliberate thought and discussion among owners and managers.

1) The **strategic**, **tactical,** and **operational plans** of the company.
2) The organizational chart (described in Chapter 1) that presents the chain of command.
3) The job title, basic duties, and qualifications of each member of the management team.
4) Customer profiles and/or the target market information about those who will buy the services and/or goods sold by the facility.
   a) This often includes customer demographics such as the age, income, location of residence.
   b) A marketing plan that explains how customers will be informed about and encouraged to purchase the services and products of the facility.
5) A description of how the facility should appear and how it is different from competitors'.
   a) This should include the location, furnishings and equipment, and customer services.
   b) A discussion about opportunities for expansion and change should be noted.
6) The financial section (discussed in Part III of the book).
   a) The business must disclose details about its balance sheet. This includes disclosures about its major assets (building and equipment), amount of cash available to operate the business, investors and any investments by the company (does it own stock of other companies), debt, and creditors.
   b) The business must also present a current income statement and project future sales, expenses, and net income. This is a budget to be created as described in Part IV of the book.
   c) Income projections should then be presented so the following financial information is known:
      i) Separation of expenses into fixed (constant) and variable (change with the level of business) so break-even can be calculated.
      ii) A profit that ensures capital equipment is replaced as it becomes old or outdated.
      iii) A profit that ensures expansion expectations can be met, such as the addition of a second service facility in the future or ownership of another company.
   d) Income and expense projections that provide monthly **targets** (expectations) for **benchmark reviews.**
7) *Contingency plans* about what will be done if benchmark reviews indicate a target has not being met.

FIGURE 3-2 Outline for a strategic business plan.

Overview of a Strategic Business Plan

As shown in Figure 3-2, the business plan contains three separate sections (strategic, tactical, and operational plans) that will act like a road map. As part of the road map, these plans answer the questions why, where, what, when, who, and how. When the questions for the plans are answered thoroughly, the owners and managers can fill in the information required for the remainder of the items (2–7) in the business plan.

As Figure 3-3 illustrates, the strategic, tactical, and operational plans must go together; however, environmental issues may cast a shadow over the tactical and operational plans. The strategic plan presents the **mission**, **goals**, and **objectives** of the company. The focus for a company is the mission of the business, which spells out *why* it exists and its purpose. Next, based on the company mission, the visionary goals declare the ambitions and future aspirations of the company by explaining *what* it intends to be like in the future. Finally, to meet mission expectations and move toward the realization of the company goals, the plan must present a set of company objectives. The objectives are concrete descriptions about *what* the management will do to fulfill the company mission to move it toward meeting its goals.

The next plan is the tactical plan. This plan identifies *what* resources (money, equipment, etc.), activities (legal approvals, advertising, etc.), and people (*who*) will be needed to implement the company's strategic plan. Then the tactical plan must propose *how* and *when* the resources will be obtained.

Finally, the owners and managers must prepare the operational plan. This plan will state *how* the business will operate. The operational plan is

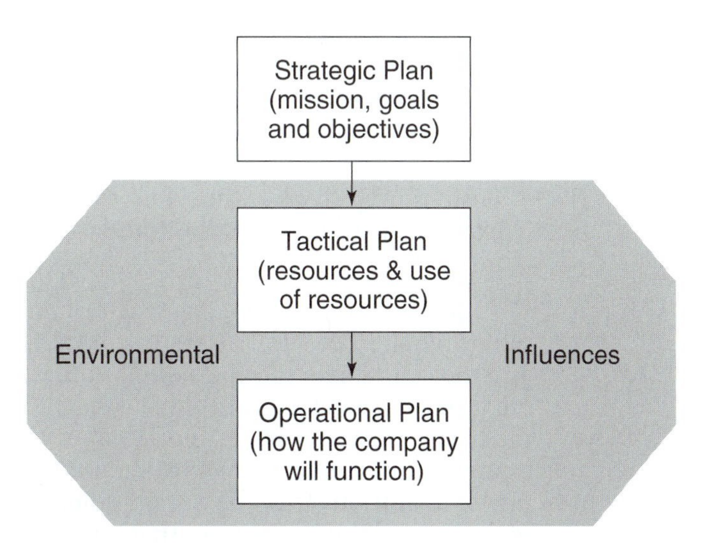

FIGURE 3-3 Strategic, tactical, operational plans.

often considered the beginning of the business plan because it contains information the service manager needs to know, which is how to run the service facility. Unfortunately, some businesses create an operational plan that does not consider the owners' intentions and long-term goals. This can cause problems, for example, resources that are not made available for the business to reach their goals financially. The operational plan not only must present the financial resources available but also must describe how the resources will be used to run the business. Of course, if the owners cannot obtain the resources, the operation plan may not be met.

In the preparation of the tactical and operational plans, environmental effects must be recognized (Figure 3-3), as there are things that are outside the control of management. For example, in one case, the city government did not permit a service facility to open because the building that was rented was zoned for a retail business and this did not include auto repair. This unique subclassification in the city code was unknown to the business owners before they rented the building, and the facility's opening day had to be postponed for one month until the zoning committee could meet and discuss the issue.

After the strategic, tactical, and operational plans are completed, responses to the rest of the items (2–7) in the business plan can be prepared. In all cases, contingencies must be prepared.

Writing the Strategic Plan

When writing a strategic plan, it must contain the company's:

- Mission—why the owners started the business (for example, to offer automobile services at a profit),
- Business goals—where the owners want the business to go over a period of time (for example, to expand from a three-bay to a six-bay shop over two years and to have all start-up debt paid off within three years),
- Business objectives—what the owners will sell to their customers (for example, maintain, diagnose, and repair all makes and models of automobiles).

These statements (see Figure 3-4) must be clearly stated and recognized by all employees. There can be three sets of separate statements, and they should be specific as well as interconnected. The mission statement in Figure 3-3 leads to the business goals. This makes sense because the business cannot have a goal that does not support the mission. For example, an automotive facility should not have a goal to become a popular radio station. This may sound ridiculous; however, all too often business owners get into ventures that have nothing to do with their mission (the reason they are in business) and this takes needed resources (money and people) away from the core business function.

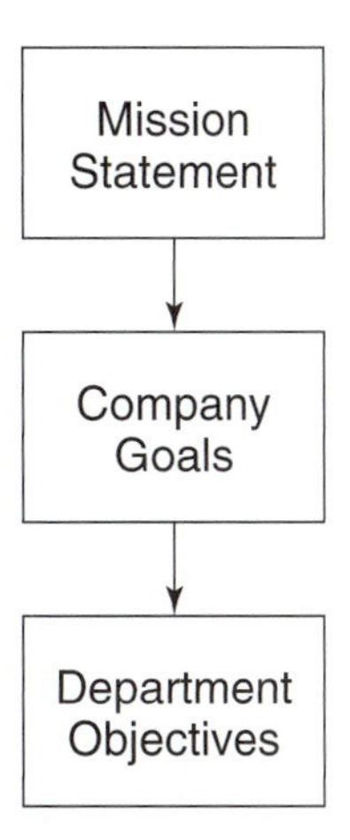

FIGURE 3-4 The strategic plan's hierarchy of statements.

For a service manager, the business objectives are important because they indicate the work that will done by each department, office, work team, and employee. With a business objective—for example, the performance of maintenance, repair, and diagnostic work on all makes and models of automobiles—a service facility will be established. This would support a mission statement that the purpose of the company was to offer automobile service at a profit. In other words, the business objectives fill in the details of the mission statements. Objectives could further indicate what will and won't be done, expected sales volumes per week, expected quality of work, and so on. When the objectives are met, the company should be moving toward the realization of its goals and fulfilling the strategic plan portion of the business plan. Next the owners would focus on the tactical and operational plans.

Preparation of Mission Statements

The mission statements for a company are prepared by or under the guidance of the owners. Because owners put up capital (money) to start a company, the primary mission of every business must be to make a profit so that investors receive a return on the money invested. When the business does not produce a profit, investors may take their money and invest it in something else. Therefore, it takes more than good intentions to keep investors' money in the business. A business that does not have enough capital will not earn enough profit and will be overtaken by better-prepared competitors.

To entice people to invest in a business, such as a service facility, a business must have other details in the mission statements (or additional statements). For example, the services to be performed and/or products to be sold should be described. The mission of a system specific service facility may be to repair transmissions. In this case, other services, such as the repair of a muffler, would not support the mission for which the business was created. This point may sound trivial, but it is important when there are requests for the facility to sell a product or add services, such as selling antifreeze at a transmission repair shop. This product or service has no relationship to the mission of the business. If the facility would consider this, it would require a change in the mission by the owners or upper managers.

To consider such a change, the first question an owner or upper manager should ask is: "If a customer buys the antifreeze product, will they ask questions and expect services that are beyond the scope, equipment, expertise, and training of the personnel?" Another important question to ask is, "If a transmission technician has to be taken away from a job (say a $2,000 transmission job to do a $30 antifreeze service), can the technician's lost time be recovered or will production and profit be reduced?" Also, the owner or upper manager should ask: "Can the money needed to purchase antifreeze equipment and inventory be better invested in the purchase of new or additional transmission equipment and inventory?" The answers to these questions are obvious. The facility will earn more money and a greater profit by having its technicians fixing tranmissions instead of filling someone's radiator with antifreeze. Working with a clear strategic mission means these questions have to be asked. This helps to keep the business on track, meaning the facility should do what it does best. A deviation from a mission must be carefully considered.

Another mission statement important to owners and investors addresses customer service. Every company should have a mission statement that their intent is to ensure customer satisfaction. This mission must be made known to all employees when they are hired. If employees are not willing to promote customer satisfaction or do not understand or appreciate the importance of this mission, they should not be hired. When an existing employee's attitude toward customers is not appropriate to this mission, service managers can simply reference the company mission statement (which should be in the company policy manual) and prescribe the appropriate level of discipline (verbal counseling with manager's notation, formal written reprimand, suspension with or without pay, or termination).

Writing Company Goals

After the mission statements are finalized, the owners should write the company goals. Goals are visionary and idealistic statements and must support the mission of the company. They should provide an insight into the way a business wants to operate and the direction in which it hopes to go. Needless to say, goals should be met after a period of time; however, this does not mean that seemingly unrealistic dreams should not be dreamed. It is important for service managers to be aware of and to appreciate company goals, even if they do appear to be unrealistic.

Examples of company goal statements may vary from one business to another; however, a typical goal for most businesses is to increase the number of customers or reach a specified level of sales. Of course, once this goal is met, a new one must be set, otherwise the business will likely become stagnant, uncompetitive in the market, and could possibly lose its market share (customers being served).

Another company goal may be to become the largest service facility in the city, while another could be to establish five additional service facilities in a region. The goal naturally should match a marketing opportunity to increase customers and earn greater profits. For example, five more locations may produce more profit than having the largest service facility in the city; however, five locations may require a management structure that the owners are not prepared to handle or want to handle. As such, the single large service facility could be the ideal goal. Although neither goal may be attained, at the very least these goals would inspire growth and improvement so that the business could remain competitive and avoid missing once-in-a-lifetime opportunities.

A goal statement could include a desire to provide community service to promote and maintain positive public relations. This would enhance the profit purpose of the company's mission through the promotion of the company name. As a result, management would want to ensure the name of the company appears in all public recognition statements. In addition, the other goal and mission statements of the company must be compatible with the public relation's goal and the service managers must work within the directives of each. For example, doing work for free on a nonprofit organization's vans would be a violation of the mission statement, and only the owner could authorize such actions.

Setting Company Objectives

Objectives state what will be accomplished. These statements must:

- Support the mission and goals of the company
- Be measurable

For example, the objective of a transmission repair facility might be to "successfully repair and, when necessary, remove, rebuild, and replace transmissions in automobiles and trucks." When an objective is reached, it can be recorded in terms of profit, time taken, or even customer satisfaction. Objectives also may include the number of repairs and/or rebuilds expected to be made per week. Another objective may include a competitive factor, such as performing the work at a price not more than competitors, or including better features, such as a warranty that is longer than competitors. In addition, if the business decides to provide related services that are within the mission statement guidelines (such as the repair of clutches), this would require that additional objectives be written or existing ones be updated.

Mission, goals, and objectives work together in the following way. When an automobile is properly repaired and the customer's expectations are satisfied, a company objective would have been met by both the service and parts managers. Likewise, a company goal is supported, which leads to an increase in sales and, therefore, the mission to make a profit and to create a satisfied customer is accomplished. Of course, if the repair was not completed properly or money was lost on the repair, the company's mission would not be accomplished. This would mean the company's objectives and the owner's goal were not met. This can be career-threatening for the service manager.

Setting Benchmarks

As explained earlier, measurable objectives are to be set by a business. Also, what is to be accomplished when the objective is met is considered a **target**, such as an increase in annual sales. Periodically, owners and managers should check to see if progress is being made toward meeting an objective's final target, such as a check on monthly or quarterly sales for comparison to the objective's final target. Therefore, these periodic targets, such as expected monthly or quarterly sales needed to meet annual sales targets, should be set when the objective's target is set. These periodic checks are referred to as **benchmarks** and so a benchmark review is to indicate if the company will meet a final objective. If a benchmark is not being met, then operations or objectives may need to be adjusted. These adjustments also may require that other changes and modifications be made. In other words, a business cannot wait until the end of the year to see if it will meet its objectives. This is too late!

Tactical Plans: Company Resources

After a strategic plan is completed, a company must prepare a **tactical plan**. The purpose of a tactical plan is to identify what resources are needed by management to meet the company's objectives (remember, objectives support the business goals and company mission). As a result, a tactical plan must focus on the resources needed and how, where, and when they will be obtained. To simplify the planning process, resources may be classified into five categories:

1. Money
2. Personnel
3. Supplies, parts, and materials to service automobiles
4. Equipment to perform the necessary services
5. Facilities to perform services.

Money

The money needed to start a business comes from owners and/or investors. As discussed in the last chapter, the source depends upon the type of ownership, for example, a proprietor or corporate stockholders. The initial capital, which often is referred to as start-up money, must be enough to launch the business and possibly justify a loan or a grant application. This is the first item of business for the tactical plan to address. How much money is needed and how much can be obtained from investors (shareholders) as compared to loans (which must be secured by assets such as property, plant, equipment, or inventory)?

The next item on the agenda for a tactical plan is to project income, expenses, and amount of cash, on both a short- and long-term basis. For example, a new business will need money for rent, equipment, inventory, and so on. At the same time, statistics show that a new company is not likely make a profit for a period of time, often years. Therefore, the tactical plan should estimate the loss and make sure there is enough money to get by.

If a business is well established, the tactical plan also must examine and project the income and expenses as well as the money needed to operate. If the business was to be sold, new owners of the facility would buy the equipment and inventory as part of the purchase. However, they would still need money to pay landlords, vendors, and so on. Regardless of the business status, the planning and projections of the amount of money available and needed is critical.

The Personnel Plan

One of the most important, and also most difficult, parts of a tactical plan is the identification of the different types of people needed to run the company as the owners intend. The service manager, if available, should play a key role in determining the type and number of employees needed by the facility. Failure to accurately identify the type of personnel needed and then recruiting them can delay tactical plans and result in lost profits. The tactical plan must consider the supply of human resources and whether the facility can find the people needed.

How much money employees will be paid should be considered. Setting wages for each position is critical to the recruitment and hiring process. If the wage paid is too low, the people hired may not be totally qualified because of the difficulty in attracting good employees. But at the same time, if the wage is too high, the facility may not be able to make a profit.

Plan for Obtaining Parts and Supplies

Another part of the tactical plan is the acquisition of supplies and parts to be resold to customers and the handling of them after receipt. The plan must identify the items that are needed and where they can be obtained. The process must compare prices, credit terms, and the distance of each vendor from the service facility. If a preferred vendor were to go out of business, then the plan must list alternatives.

Equipment and Facilities Plan

The tactical plan must identify the building and equipment needs of a service facility. This plan should include a short- as well as a long-term plan that corresponds to the company's goals and objectives. In other words, if the business plan proposes that after three years in operation the facility intends to offer automobile alignment services, the equipment plan must include the purchase of an alignment machine. Furthermore, the plan must indicate where the equipment will be placed in the building and if building changes are needed, for example, the addition of electrical service, removal of a wall, or purchase of a special lift. The financial section of the plan (noted earlier) would need to indicate how the business intends to pay for the machine as well as other changes, such as construction costs.

Equipment (and tools) to be purchased to repair automobiles and run the business are called assets. When the business owners also own the building, that would be an asset (although many owners will lease the building or own the building under a separate business for various legal and financial reasons). The tactical plans should identify all assets owned or needed by the facility. Being aware of these assets is critical to the

FIGURE 3-5 Equipment to be purchased to repair automobiles and run the business are called assets.

FIGURE 3-6 If a computerized system is going to be used for the parts specialist the tactical plan must insure the software and technical support is available.

service manager when preparing the operational plans. For example, if a service facility is going to use a computerized system for the parts specialist to order parts, the plan must ensure the software and technical support is available. The plan must include how the computer will be obtained, placed, and installed (for example, will wireless, cable, or telephone modem be used). Also the plan must indicate when and how its effectiveness and efficiency will be reviewed.

In addition, the plan must address the facilities. Questions to answer include: Will the building housing the business be large enough for the current operations or too large? Will the size of the building permit the business to expand in the future?

In one case, a repair facility moved into a very large building that used to be a dealership. The owners were sure that the facility would never need all of the space, and so they rented a large part of it to a factory for storage. They were wrong. Their business took off and within a year they were out of space. The point of this example is that owners and managers must make thoughtful and considerate decisions when preparing the business plan. Obviously, these owners did not consider any long-term goals and this cost them profits.

FIGURE 3-7 The tactical plan must address the type of building needed and the optimum size.

Public Relations Plan

The purpose of the public relations as part of the tactical plan is to ensure that the business will have enough customers to make the break-even point. To promote the business, the plan must determine the mix of advertising sources, including newspapers, radio, television, brochures, flyers, and so on. It must indicate which of these advertising sources will be used on a monthly basis and which will be used quarterly or seasonally. Then a variety of other questions must be asked, including: What types of advertising will not be used? Should the amount spent on advertising be a percentage of sales or a fixed (constant) cost each month? Will the methods used to advertise be reduced after sales increase to a specified level? If sales dip below a specified level, will the advertising increase? Should advertising be increased in response to a sales dip or in anticipation of a sales dip? Will the company place advertisements with special-interest or community groups that are raising money? If so, what criteria will be used to determine the amount or type?

These and other questions must be asked before and when a facility is in business. After the plan is laid out, some employees, such as the service consultant, must be informed about the advertisements (such as an oil change special in a flyer sent to regular customers). An operational report must then be prepared on a monthly and quarterly basis with reviews based on data that indicate whether the advertising was successful (for example, did gross sales increase; if so, did net profit also increase relative to the cost of the advertising). This review is important and all managers should be included.

Operational Plan

The operational plan is critical because it describes how a company will function or work. The plan has two purposes: first, as discussed in Chapter 1, a business should describe how resources will be used to achieve the company's mission, goals, and objectives. Second, the plan must be followed when preparing the operational procedures for the delivery of the services and products to the customers. In turn, the operational procedures are used to prepare the operation manuals to which the employees can refer.

One approach to preparing an operational plan is to ask and answer questions. Several of these questions would be:

- What are the responsibilities and authorities of the different levels in the organizational diagram, for example, the president or owner level, vice president or division management level, the department management level, and so on?
- What are the responsibilities and authorities of the company officials or managers at each level in each unit (a unit may be an office, division, department, or recognized group within a unit such as a team)?
- Who reports to whom, and how?
- What resources will be provided to each unit, for example, office space, furniture, equipment, staff support, and so on?
- How will budgets be prepared, submitted, reviewed, approved, and monitored for each unit or department?
- How are repair orders processed?
- How will employees be hired, trained, and evaluated at each level?
- Who will supervise employee evaluations?
- When and how will unit reports be prepared and evaluated?

Clearly, there are many other questions to be asked. These questions will be different for each company, however, because of size (large versus small organization), method of ownership (proprietor versus a C Corporation), and type of business (independent repair shop versus a dealership). This is also the case for the operational procedures. In describing how work will flow through a facility, a diagram with a set of directions to explain the actions to be taken in each step of the diagram (which will be presented in the next chapter) should be used. Obviously, the diagram, procedures, and operations manuals must be custom-made for the service facility. This is difficult, or even impossible, if an operational plan does not exist.

The Environment

One of the reasons that one business plan cannot fit all facilities is because of the environmental influences in which they must operate. Businesses exist in an **open business environment**, which means there are various forces and events outside of a manager's control that affect a service facility's operation. A simple example is snow (shown in Figure 3-8). A service facility in the north must face this type of weather each season, but in Hawaii, service bays can operate without garage doors year-round. Service managers must recognize these outside influences to take advantage of their positive features and to counteract their negative influences.

Two business environments that are particularly important at a service facility are the **tactical** (how and when resources needed to support the strategic plan are obtained) and the **operational** (how the business operates) environments. These environments influence the tactical and operational plans. It is important for a service manager to recognize the differences when going from one service facility to another. What worked in one place may not work at another because of the environment. For example, washing cars after a service may have been a desirable feature for customers at one service facility but results in wasted time and customer irritation at another. In other words, customer expectations as well as the availability of a wash bay sets the operational environment for the facility.

The tactical environment includes the influences that directly affect the business resources (personnel, capital, and equipment) needed to conduct daily business operations. For example, service managers of automotive facilities need personnel (technicians), tools, and equipment, as well as parts to conduct business. If any of these resources are not available, the ability of the facility employees to do their job will be limited, which, in turn, will limit sales. The pace at many service facilities is fast. Customers want their services delivered promptly or the sale will be lost. Therefore, parts must be delivered quickly and cannot take a day or week to get to the facility. If the vendor closest to the facility does not carry a full line of parts and the one that does is over the mountain, the tactical environment is negatively affected. This means that the section in the tactical plan that covers availability of parts must note this limitation as a negative impact. The plan must then seek a remedy to the problem or reduce sales projections.

The operational environment represents influences that affect day-to-day business activities (operational plans). The influences on operational activities include local demographics (such as population and age), economics (such as average income or high unemployment), and sociocultural features of a community (such as educational level and population mobility). Each of these will affect what customers expect from the service facility. The operational environment also may be effected by fea-

A

B

FIGURE 3-8 An open business environment means there are various forces and events outside of a manager's control that affect a service facility's operation. An example of this is a) the harsh weather of the north or b) tropics of Hawaii.

tures such as the location of the facility, for instance, a building on a main street near the center of town versus a side street in a residential setting. For example, a facility in a community with a very high average income may have a large percentage of people who own or lease new automobiles with warranty contracts. In response, an independent service facility may

FIGURE 3-9 It is important that the facility appeal to every customer, added features such as vending machines help ensure the customer is comfortable.

need to engage in special advertising to offer maintenance specials with discounts or even add new services (such as a car wash or quick lube station). This may pose an opportunity or a threat and could change the company mission statement.

Talented service managers must attempt to control the operational environment's effect on the service facility's daily operations. However, problems with the tactical environment are more difficult to counteract and often require extensive resources to fix. For example, service managers may make the service facility's property appeal to the people it serves and keep the property clean, attractive, and landscaped. In addition, the manager may make extra efforts to keep the building interior clean with a relaxing, cheery waiting area, possibly with toys and children's books as well as snack and soda machine (as shown in Figure 3-9). The entire property must be made to appeal to every customer (female, male, young, old) and not just to the owner, workers, or a small subset of the customer base. These changes will cost money; however, some changes will cost more than others and some will be free. For example, old parts, new parts, cores, and used parts should never be left lying around where customers can see them. This can make customers feel uneasy when they are at a service facility. In general, making improvements will often help public relations, which typically will improve sales.

Concluding Thoughts

Although preparing the above plans is important, their review and update is just as important. All too often, owners and managers avoid or give a low priority to working with their plans. They may believe the daily operations are more important. The reality is they avoid this review because the process requires decisions to be made and decision making is not a pleasant task. It is unpleasant because there is no correct decision but only a *best decision*. Specifically, committing the business to an action that is not guaranteed to be 100 percent correct is difficult. As a result, the review of the business plans, procedures, and operation manuals is often avoided. To do nothing is actually a decision, one that may lead to failure.

Another reason for avoiding the reviews of plans is because managers would rather get on with operational activities. They are anxious to get customers into the facility, get parts on cars, and make a profit. This is not bad, but it is hardly an excuse not to review plans and procedures. Unfortunately, neglecting how a facility does business is a reason why customers and employees complain that things never get better. Unfortunately for the managers, owners, and technicians, customers know when things are not done right. Therefore, the failure to review and modify plans and procedures to make the service department better is considered a failure in leadership, not necessarily management. Customers and owners may feel that the reason something is not done properly is that the manager cannot get the employees to do the work properly. However, the reality is that the employees are doing the work properly as per the operational procedures. These procedures are simply too old and inefficient, and maybe even not in line with the practices of other businesses.

To put planning in perspective, compare it to preparing a room for painting. Removing old paint from the walls and woodwork, sanding the rough spots, masking where paint lines are to end, covering the floor, and so on is work most people do not like to do. They would rather put paint on the wall and enjoy the immediate improvement to the room. Failing to prepare a room properly before painting, however, means that the rough spots in the surface will show up after the paint is applied, paint may not stick to the wall or doorway, paint lines will not be straight, paint drips will be seen on the floor, and so on. Likewise, when time is not taken to prepare and review company plans, the rough spots will show, operations may be sloppy, and company profits will not be maximized. It is much the same with respect to planning as a managerial responsibility. Plans must be reviewed on a regularly scheduled basis with input from as many sources as possible.

Review Questions

1. Outline the main parts of a strategic business plan.
2. Describe the purpose of strategic, tactical, and operational plans and how a manager will work with each one.
3. Describe how to prepare mission, goal, and objective statements for a service facility and give an example of each.
4. Describe how operational goals and objectives for a service facility are prepared and provide a written example.
5. Describe why a management budget is important and how it should fit within the tactical and operational plans.
6. List the different types of environmental influences that can affect a service facility in your town or city. Then explain how each affects the tactical and operational environment as well as how the strategic business plan might change.

CHAPTER 4

PREPARING FORMATTED SYSTEMS

LEARNING OBJECTIVES

Upon reading this chapter, students should be able to:

- *Describe the purpose of:*
  - *company policy*
  - *operational procedures*
  - *an operation manual*
- *Explain why company policies and the operational plan are connected to operational procedures and operations manuals.*
- *Describe the advantages of a formatted system.*
- *Describe the stages of a system and the importance of feedback.*

Introduction

The previous chapters introduced and described organizational structures, types of business ownerships, types of automobile facilities, and the strategic business plan. Basically, these topics addressed the bigger picture. Another piece to be added to the big picture is company *policy*. After owners and managers complete the big picture, company policy must be connected to management's day-to-day activities through the preparation of the operation manual. System franchises refer to this as a formatted business system because it connects the big picture to procedures and documents used every day by managers and employees. Therefore, the discussion of the big picture connection to the managerial documents is the purpose of this chapter.

As explained in the last chapter, the company strategy includes an operational plan that describes how the strategy (mission, goals, and objectives) will be achieved. The operational plan is then used to prepare **operational procedures**, which, in turn, are the basis for the creation of the **operation manual**. The operation manual contains rules and regulations and may also be known as the service department procedure handbook, employee handbook, work area manual, or even the company bible. Each one's purpose is the same and employees must learn to follow it closely.

Operational procedures are established from management's operational plan (discussed in Chapter 3) and company *policies* (see Figure 4-1). At unionized service facilities, some operational procedures on topics such as employee discipline, minimum performance expectations, and complaints (called grievances) may be directed to the union contract. Therefore, managers must be aware that company policies are used to prepare operational procedures and operation manuals. Consequently, company policies are not secret documents but, rather, tell employees what their employer expects from them. They are codes of conduct that follow rules as well as good business practices. Policies also tell employees what they can expect from employers (such as vacation time, benefits, and review of their performance). Therefore, policies also make up the rules and regulations of a company. This is why a rule or regulation may be referred to as a company policy.

Making Company Policy

The owners (corporate board of directors along with the corporate officers) are responsible for establishing **company policies**, which will protect the company and direct the conduct of the company employees. Managers are expected to enforce corporate policies through oper-

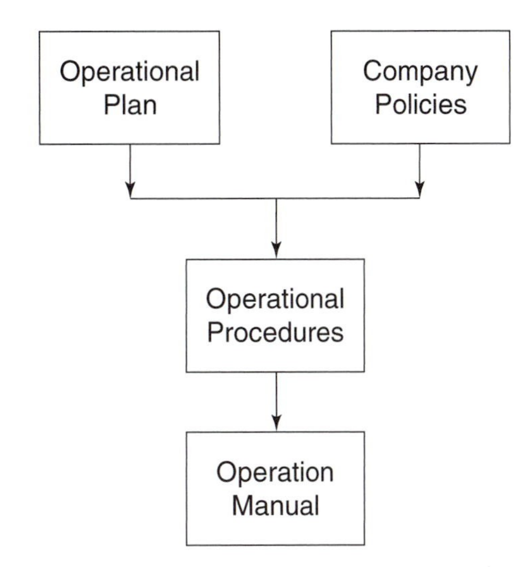

FIGURE 4-1 Flow of operations planning/procedures.

ational procedures. In turn, the procedures are used to create the rules and regulations (operation manual) for employees to follow. In some cases, company policy may be so specific (such as a dress code that might require hair to be short or tied back so it does not pose a safety concern) that it will become a rule in the operation manual. Another example of a company policy is the hours of business; for example, it may state that the business will open promptly 8:00 A.M. from Monday through Saturday except on national holidays. This policy would simply be repeated in the procedures and operation manuals of each unit.

Here is an example of a company policy that becomes an operational procedure and then is included in the operation manual:

1. Company policy:
   A. Due to liability concerns, all parts put on a customer's automobile must be supplied by the service facility.
2. Operational procedures:
   A. All parts needed for customer automobiles must be ordered by the parts specialist.
3. Operation manual:
   A. When the service consultant has an authorization to repair a customer automobile, all parts will be ordered by the parts specialist.
      i. Customers MAY NOT bring parts to the service facility for installation.

Company policies also come from public law. Some of the legal policies are quite complicated, including:

- Federal laws on employment, sexual harassment, discrimination, safety (OSHA), providing a drug-free workplace, right-to-know (EPA), unemployment compensation benefits, insurance coverage, and warranties.

- State laws on safety inspection guidelines, **lemon laws** (especially important for new car dealers), mechanic liens (prohibit a service facility from holding a customer's car for payment unless a certain steps are followed), consumer protection (advertising disclosures and how to legally interact with the general public), as well as any state employment laws.
- Local laws such as the use of properties for the service of automobiles, noise ordinances, parking requirements, waste removal and management, and so on.

In other words, all federal, state, and local laws are guiding principles that direct the conduct of the company and its employees. These laws must become part of company policy and company policy should never violate any of these laws.

Managers may add to the policies of a company, but the owners should approve them in order to carry out their authority. Of course, because a service manager is an expert in automotive services, he or she should help to prepare or carefully review the operational procedures and company policy. In addition, when a new service manager takes over, he/she should review company policies and possibly modify the operational procedures and even the operation manual to bring his/her expertise into the facility.

The creation of company policies is a lot of work, but it is necessary to protect the owners and resources of the company. As a result, a novice should not prepare them. Instead, a company that specializes in the creation of policies or a human resource specialist with the assistance of an attorney should prepare policies and a policy manual. When a policy is created, it must be stated properly in order to comply with legal expectations. Each policy must indicate its source, such as an owner's action with the date of the action, a law with a legal citation, or a manager's directive with the initials of the manager and the date of owner approval. In addition, a policy manual should be thoughtfully prepared by dividing it into sections with a numbered outline. Placing directives and laws in a filing cabinet does not substitute for a policy manual and will not work when problems occur.

Because of the legal importance of the policies, a policy manual should be placed in an office where employees have access to it. Of course, more recently company policies can be made available via computer files or Web pages. Regardless of the method of compiling company policies, all new employees should be given company time to read them. After the policies have been read, the employee should be asked if an explanation is required. After an employee indicates he or she had read and understands the policies, he/she should sign a form that states the person has read and understands them. The form should then be placed in the employee's personnel file with a copy in a company file. This record is important for a number of reasons. For example, assume that an employee sexually harasses another employee. Among other things, the company must show that the employee was informed of and understood the company policy that opposes such behavior.

As new policies are added to the manual, the operational procedures and operation manual must be updated. In addition, all employees must be informed of the changes in policy, procedures, rules, or regulations. This may require that each person is given a hard copy to read and signs a form indicating he or she has received it; however, in some companies, employees may be informed by e-mail.

The Operational Plan

As explained in the last chapter, the operational plan is prepared after the strategic and tactical plans are completed. The operational plan describes the ways in which the business will function or work to meet the expectations presented in the mission statements, goals, and objectives. The operational plan also must provide the information needed to create operational procedures for the company and an operation manual for each unit in the facility. Although some of the rules and regulations in the operation manuals will be the same, some will be different. Of course, the procedures, rules, and regulations must go together to meet the course of action expected in the operational plan. Ensuring that the operation manuals are in agreement and relate to the operational plan is part of the operational procedures.

The Connection: From Strategy to Delivery of Services

Therefore, the operational plan must have a direct relationship to the strategic plan, on the one hand, and to the operation manual, on the other. When this occurs, the strategy will be aligned with the day-to-day activities so that the business will head in the right direction. The managers are like the drivers of the automobile; they can point the car (business) in the wrong direction and run off the road.

Operational Procedures

With an operation plan in place and a copy of the company policies in hand, the operational procedures can be prepared. The primary purpose of operational procedures is to ensure that each unit knows what to do so that the work will flow smoothly. As a result, the operational procedures must have at least three parts:

1. The identification of departments, offices, or groups of employees (referred to as units) who are included in the procedures. For example,

there may be a service department, parts department, and cashier's office.

2. Objectives and performance expectations for each unit must be declared. For example, what are they expected to do and how well are they to do it?
3. Work flow between the units, such as the flow of repair orders, computer entries, ordering and delivery of parts, payment of invoices and delivery of customer automobiles, as well as procedures for the maintenance of equipment, hiring, evaluations, collecting and depositing money, and so on.

Unit Objectives and Performance Expectations

After the units are identified, the procedures must indicate what is to be accomplished by each unit and how and when it will be determined if the performance was acceptable. In other words, the unit objectives are the company's objectives taken to a practical level that is appropriate for employees. Therefore, unit objectives need to directly support company objectives, which are tied to the company's strategic goals and mission.

Because objectives state what will be accomplished, they should be written in measurable terms so that performance targets can be identified and checked. In identifying performance targets, there are two types of targets to measure. One measure concerns **effectiveness**, which regards *outcome* targets, and is an assessment of quality performances. For example, **comebacks** (also referred to as *second attempts*) occur when a customer had a repair made to an automobile and must return to have the same repair made again. The need to have a second attempt to repair the same problem indicates that the initial work was not effectively done, and the company objective to perform quality work would not have been met for this job. The damage to the company's profit would depend on the company resources (technician time, parts, and supplies) needed to make the repair a second time.

The other measure regards **efficiency,** which is assessed in terms of *output* targets and focuses on quantity production. Outputs look at the amount of work performed regardless of quality. An example of an efficiency target would be the number of sales or amount of money earned each the month. In such a case, the target for the objective may be to increase sales or to increase sales by 10 percent. This type of efficiency measure cannot continue to be positive because sooner or later either the amount of work produced will max out or quality will go down. A better measure would be to compare the increase in sales to the number of hours worked by the technicians in the service department. As the amount of money earned goes up and number of hours worked by technicians remains the same or even decreases, the result is greater efficiency. Conceptually, the idea would be to reach equilibrium, where quality and quantity are in balance.

Obviously, problems occur when services become *too effective* to the point that efficiency is lost. (For example, each job is checked and rechecked over and over again to the point that excessive employee time and therefore money is lost.) Problems also can occur when services become *too efficient* to the point where effectiveness is lost and comebacks (second attempts) occur. In both cases, a service manager will have to review employee work performances and possibly reorganize the department, make reassignments, create additional targets, or even revise existing targets.

Work Flow Procedures: A Diagram and Narrative

The presentation of the **work flow process** begins with a diagram, such as the one presented in Figure 4-2. (Note: This diagram is for illustration and discussion here. It does not attempt to present an actual diagram of all of the actions and activities in the process.) As this illustration shows, the work conducted (a section of the delivery of a service) flows from one unit or person in a unit to another. Basically, the diagram is like a road map for a work order as it travels from box to box in the diagram. Each box represents an activity or action in the process.

Obviously, the diagram in Figure 4-2 includes every possible step or tells the entire story. The figure begins with the signing of a work order by a customer and ends with the payment of the invoice. In the diagram, three different units or departments could be involved: the service department, the parts department, and the business office. The details for the procedures for each unit are written down in its operation manual, such as a manual for the service department, parts department, and cashier's office. For example, the written information for the operational procedures for a diagram as shown in Figure 4-2 would indicate:

- Who (what unit and possibly the position/person in the unit) prepares work orders (Box #1).
- Who presents the invoice and what is to occur when the customer signs the work order (Box #2).
- When, how, and who delivers work orders to team leaders (Box #3).
- When and who (the service consultant, team leader, or technician) gives the parts department the repair order to pull the parts (Box #3a).
- Who delivers parts and how they are delivered to the technician (Box 3b).
- Who assigned repair orders and how they are assigned and delivered to technicians (Box # 4).
- How and from whom (unit and possibly person) does the service consultant receive the repair order (Box #5).
- How and from whom (unit and possibly person) does the service consultant receive the parts information (Box #6).
- Who creates and who files the invoices/repair orders (Box #7).
- When and to whom (unit and person) does the customer pay the invoice (Box #8).

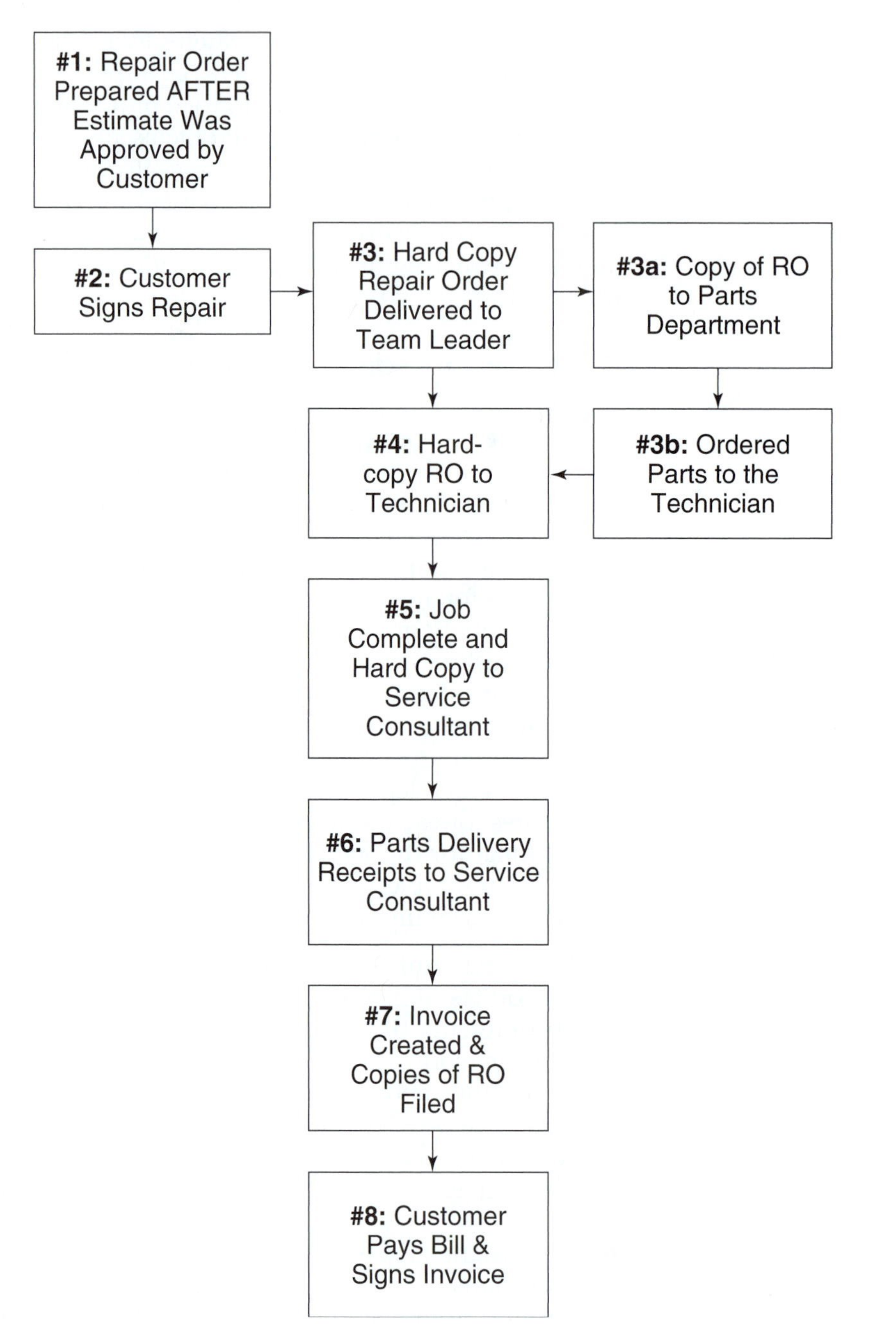

FIGURE 4-2 Flow of repair order for a job with preordered parts.

Operation Manuals: Creating a Formatted System

Using the operational procedures as a guide, the operation manual for the service facility and those of the other units break down the diagram and description of the procedures to explain what the unit and the employees in each unit do. The manual is not a job description (described in Chapter 5) or set of job descriptions but, rather, a document intended to obtain a consistency in the performances of the employees in the unit. Often the process created by the operation manual is referred to as a **formatted system**.

To create a formatted system, the manual must describe in detail the activities to be performed by a unit on a day-to-day, week-to-week, and month-to-month basis. For example, the manual should include instructions on activities such as:

- Working with customers (from greeting to closing the sale)
- Safety procedures (personal safety, safety of others in the shop, safe equipment use, and reduction of breakage of parts/damage to customer property)
- Handling of customer automobiles (placement of paper mats and plastic seat covers)
- Work flow (how the repair order is processed, who is responsible for what specific aspect, and who has the authority to make changes/decisions)
- Standard service procedures (some repetitive jobs may require standardized procedures such as how to conduct an oil change, inspect a car, preliminary engine diagnosis work, battery inspection, among other operations)
- Opening and closing the facility (checking each work area, what lights to turn off and on, setting the thermostat)
- Checking inventory (checking available oil, filters, and other parts for sale) and placing orders for supplies (brake clean, shop rags, etc.)
- Servicing shop equipment (maintenance of the shop air compressor and lifts), and so on.

Obviously, some items in a manual apply to everyone in the service department, whereas others apply only to a smaller unit or group of employees. Consequently, there are usually several operation manuals. For instance, there could be a manual for the service department and one for the parts department, while in the service department there may be a manual for technicians to follow and another for service consultants to follow. The instructions that are common to everyone may appear in all of the manuals, such as how to act when approached by a customer or acceptable places to take breaks. However, at some businesses, common information (particularly information about employee behavior and conduct) may appear in a company policy manual. In this case, the operation manual would just refer to the policy manual.

Because the operation manual contains work area–related information, some of the instructions in the manual become performance standards and are used by managers to assess employee performance. For example, an operation manual may indicate how information is to be entered into a computer, where work orders are placed, where to file work orders, and where final invoices are placed after a service is completed. If an employee cannot enter data correctly or fails to put work orders or invoices in the correct place, then the efficiency and possibly the effectiveness of the facility will likely be harmed. In other words, time is wasted when an employee or employees have to find a misplaced work order. Time costs a company money. In such cases, an employee may need more training. If additional training is not relevant to the failure in performance, a warning should be issued and possible termination or reassignment may be necessary (this will be discussed in the next part of this book).

If changes in operational procedures or the work process are made (such as the processing of work orders from the parts to the service department or the placement of the work orders in the shop), the service manager would inform the employees affected and change the operation manual. As a result, understanding the use and relationships of the operational procedures and operation manual is important to service managers. In some cases, a service manager may want to discuss a change in practice before implementing it. Because the operation manual should describe current practices, employees can offer an informed opinion. If the employees do not have a firm grasp on current procedures because an operation manual does not exist, they can only make suggestions based on their limited work experience in the facility.

Franchises and chains believe that the recognition of consistency by customers improves business. This is because consistency gives customers confidence that, regardless of the location where they are served or who provides the service, they will receive the same services and workmanship. This is the reason that most national franchise corporations require local owners to use formatted systems for operations. Usually, a franchise owner is given or required to purchase the training needed to implement the system as well as equipment and any needed computer software to manage it. In other words, a service manager of a local franchise receives a proven set of operation manuals (or formatted system) that serves as a cookbook for conducting day-to-day activities.

Some formatted systems developed by a franchise have set industry standards. One of the best-known cases is the work of Ray Kroc, who founded McDonald's. Kroc wanted to assure customers that they would get what they expected at every McDonald's franchise around the world. Because of Kroc's success, his procedures were copied by other businesses. At the same time, however, Kroc's approach showed the benefits gained when a company's business plan and policies are aligned with day-to-day operations. Local owners could stay on track by following the formatted system.

Preparing and Reviewing the Operation Manual

The contents of an operation manual for a service department should cover its daily activities. The service manager can prepare and review an operation manual by simply walking through the daily activities performed at the facility as well as examining weekly and monthly activities performed by the manager to maintain business operations. For example, a service department's operation manual for a service consultant should include procedures and policies for the following:

- Information about appropriate dress and appearance for the job performed
- Telephone procedures from talking to customers to personal phone calls
- Information about greeting customers, from what to say to how to act
- Information on how to schedule appointments, inform customers of estimates, and obtain approvals for repairs
- Information on using the computer for activities such as
  - Entering and maintaining a customer log
  - Entering, maintaining, and using customer files
  - Preparing estimates, work orders, and invoices
  - Ordering parts
  - E-mail communications
- Information on distribution and processing of work orders
- Information on care of customer automobiles by the use of fender covers, seat covers, floor mats, and steering wheel covers
- Information on ordering parts and details about special-order parts
- Delivery of parts to technicians or notification of parts arrival to the lead technician
- How to write notations on repairs
- How to keep a daily log on the status of services
- How to keep customers informed
- How to present an invoice and car keys to customers
- How to conduct pre- and postinspections of customer automobiles
- Plus information that would depend on local conditions and owner expectations

Systems Management

When a service is provided, the work required to provide the service as described in the operational procedures and operation manual should be separated into *stages* and put into an order. When the stages fit together properly, they create a *system*. When the work to be done within the system is understood by employees as a series of sequential stages, managers can more easily identify strengths and weaknesses of the department. This will mean an encountered problem can be thought of as an opportunity to improve effectiveness and/or increase efficiency of the people working in the system or the system itself. Of course, the systemization of work does not guarantee success (improved effectiveness or efficiency) but, rather, is an attempt to organize the work in order to maximize results.

System Stages

A basic system must have at least four stages. As shown in Figure 4-3, a system must at least have an *input*, a *process*, an *output*, and *feedback*. Although a system can be expanded and divided into subparts, the stages are still the same. In order for a unit to be productive, these stages must be in harmony.

As shown in Figure 4-3, the first stage of a system is the input stage. This is where the work comes into the unit. What comes in and how it comes in are the two concerns for a manager. What comes in depends on the type of services the facility offers (the mission of the service facility). If the service facility repairs transmissions, then transmission work (not engine work) is an acceptable input. If it does not repair transmissions, then a transmission repair should not be accepted.

One may believe that controlling the input into a system is simple; however, this often is not the case. For example, a job taken into a repair facility cannot be handled by any of the technicians because it is a drivability problem in an unusual imported make and model of automobile. When this occurs, an embarrassing situation is created and a lot of time must be taken to resolve it. Possibly the automobile can be returned to the customer, which creates a public relations problem. Another option is that the work be taken to another facility where the repair can be made, but this means the business will not make a profit on the repair. Either solution creates a loss, as employees try to help the customer while also accomplishing the company's mission. The end result is that it will cost the company money. The service manager must correct this problem at the input stage.

How the input stage is handled by the facility is critical. Assume that a customer comes to a facility with a service problem. Before it can be diagnosed for repair, proper procedures must be followed by the service consultant. The service consultant must properly document the problem

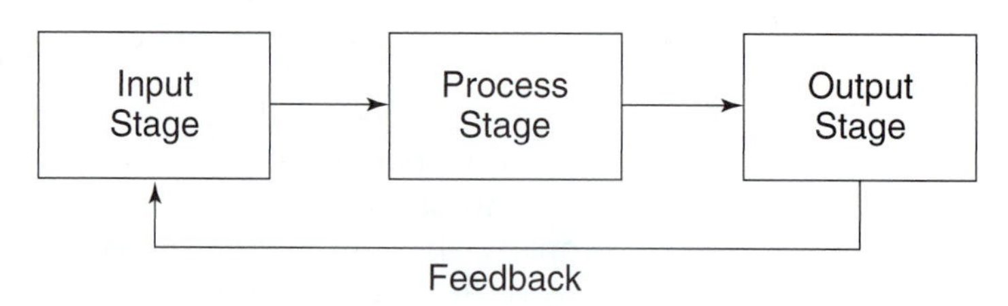

FIGURE 4-3 Three stages of a system.

and prepare the necessary paperwork before the job can be taken to the process stage. If this part of the input stage is not done correctly, a sale may be lost (the customer is not served properly and leaves), a diagnosis may take extra time or not be conducted properly (because of inadequate information collected by the service consultant), a maintenance may not be performed (the service consultant neglected to ask questions), a product may not be sold (the service consultant fails to tell the customer about various options), and so on. In other words, an input stage should have a system within the system and if it does not work, it will create confusion and cost the facility money.

After the input stage, the service moves into the process stage. At this stage, the resources (technicians, equipment, repair space, parts and supplies, staff personnel, and so on) are used to provide the maintenance, diagnosis, or repair. In this stage, the service manager must ensure that all resources are utilized effectively and efficiently.

At some facilities, the system design may have to be modified to fit the different type of work done at a service facility. In Figure 4-4, the modification is to the process stage in which there are differences for maintenance work, diagnosis work, and repair work. As shown in Figure 4-4, it can be assumed that:

- At the input stage (Figure 4-3), the customer makes a request to the service consultant and the service consultant must select the proper process to use (this will dictate what technician can be assigned the job, in what order the work can be done, and how long it will take).
- The job goes to the process stage (Figure 4-4) for a diagnosis.
- A technician finds the cause of the problem and a repair is recommended. The work order is returned (note that the diagram shows that the input to diagnosis arrow has two points because work can flow

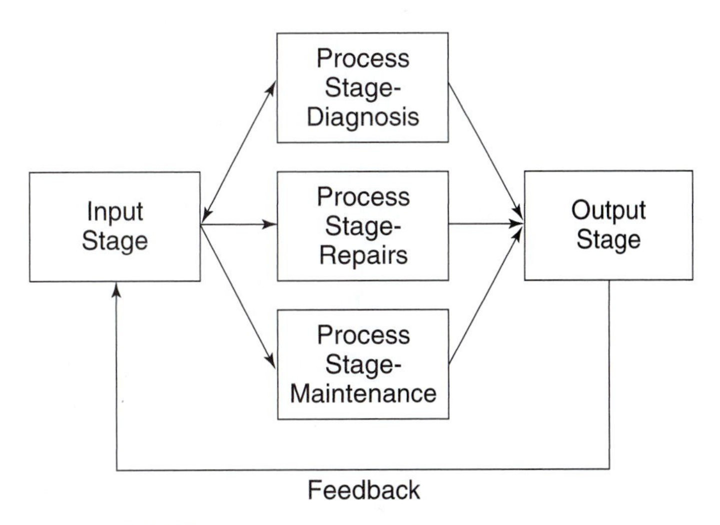

FIGURE 4-4 A system with three process stages.

both ways) to the input stage in order for the service consultant to obtain the customer's approval to make the repair.

- If the customer approves the repair, the work order must then go back to the process stage for a repair. If not, then the diagnosis is complete and the repair order will go to the output stage.

As shown in Figure 4-4, the design of the system for a service department includes three different process stages: diagnosis, repair, and maintenance. Each process stage will require a different procedure for the processing of customers' requests on repair orders. Each of the process stages must have operational procedures created by management with the input from employees who will use the system when possible. Ultimately, the operation manual will contain the procedures and rules the employees will follow and management must train the employees to follow the procedures.

Finally, the process stage will also depend on the ability of the technicians to do the work. For each of the process stages, the technicians also will have procedures found in the operation manual to follow. These procedures may be step-by-step instructions with technician check-off sheets to fill out (for example, a safety inspection). The procedures may be more general, such as the manufacturer's service manual procedures that are to be used with specifications and actual measurements recorded on the repair order. In addition, a service facility may have a procedure that requires lower-level and lower-paid technicians to do certain types of work (or parts of work) such as hook up diagnostic equipment, remove trouble codes, and look up needed information. A higher-level and higher-paid technician may be used to perform the actual diagnosis, because it may become quite complex. Furthermore, technician procedures may address the availability of equipment to do the job, space, and parts/materials constraints, and the directive for employees to focus on the completion of the job. Of course, accidents happen, equipment breaks, parts do not show up on time, people get sick, and so on. Consequently, the process stage or stages at an automotive facility often require contingency plans and attention of service management to intervene when problems arise.

After the work is completed (the service has been performed on the automobile), the product is ready to be picked up by the customer and the job moves to the output stage. At this stage, the invoice must be prepared, payment must be made (possibly credit arrangements must be made), and the automobile must be properly delivered to the customer. The recognition of the move from the process to the output stage is important to the service manager's ability to solve problems, and improve effectiveness and efficiency.

After the work in the output stage is completed, the feedback stage must be addressed. All too often, formal feedback is not conducted because employees are too involved in the input, process, and output stages. Feedback, however, is needed to assess how the stages are working, and if the service met both the customer's expectations (price and performance) and the owners' expectations (profit and adherence to their business plan). A service manager should use both formal and informal means to obtain feedback on a job. The manager must seek out information on:

- Customer concerns (monitor comebacks and second attempts)
- Satisfaction (discussed in greater detail in the *Service Consultant: Working in a Service Facility*)
- Repair orders to determine whether of not proper labor times were charged as compared to a labor time guide
- Parts markups
- Gross profit realized when the amount of money charged for the technician's labor was compared to the cost of the technician's labor

In addition, the manager must examine work performances at each of the different stages to determine whether or not adjustments should be made to improve performances.

Managing a System at an Automotive Facility

A service manager must *plan, organize, control,* and *lead* in order to be in command of a service department's operations. Without question, using a system approach to management in order to meet these responsibilities is challenging but not as difficult as meeting them without a system. Without a system, a service manager must attempt to keep track of a lot of details that may not appear to be tied to each other in order to effectively and efficiently meet the expectations of customers and owners. Basically, the use of a system's approach is why some service managers are in command of a growing service facility that accomplishes the company's mission and helps the owners meet their goals. At the same time, other managers may be in a constant crisis because they attempt to manage every day-to-day detail and aren't really meeting a strategic business plan.

To start to understand how to manage a system at a service facility, Figure 4-5 illustrates how a typical automotive service facility provides maintenance, diagnostic, and repair services. To provide these services, the service manager must recognize that at a minimum three different classifications of employees must work together (technicians, service consultants, and parts specialists), and at some facilities more classifications may exist (such as a lead technician, cashier, and so on).

During the day, customers request services that require different technician skill levels because the work varies from job to job. Specifically, low-tech services often take little time to perform and relatively little expertise to complete successfully (such as an oil change). For these jobs, someone with a limited knowledge of tools and equipment can be trained

FIGURE 4-5 Different technician skills and services make the job of scheduling and assigning work to maximize gross profit a challenge.

in a relatively short period of time. At the same time, most service facilities (except those that only provide low-tech maintenance services) perform high-tech repairs that take hours to complete (such as a transmission overhaul), require expensive tools and equipment (such as alignment machines), and need technicians with extensive training and experience (such as computer-controlled diagnosis and repair). This makes the job of scheduling and assigning work to maximize gross profit challenging.

Consider that, within this complex setting, other employees with different job skills such as a service consultant, parts specialist, and even a cashier, are needed to work in the input and output stages. Truthfully, the only way for a service manager to plan, organize, lead, and control the operations in a department composed of the above people is through systems management of each stage. This means the creation of a comprehensive operation manual that has clear procedures for as many of the job duties each employee must perform as possible. Without it, the only option is to get smaller and to personally perform as many duties as possible yourself.

To create a system for a service facility, the service facility first must have operational procedures and company policies already in place. The procedures and descriptions of activities to be conducted must be interconnected and placed in an operation manual or manuals (there could be one for each classification of employee). This is the part of the service manager's job that requires *planning* and *organizing*.

FIGURE 4-6 A service manager must use various methods to track operational procedures.

Next, the service manager must use various methods to track these activities through the system. By checking the performances of employees relative to the feedback stage examined above, the service manager is able to *control* operations. If the planning, organizing, and controlling of operations and personnel performances are done correctly, the service manager will *lead* the facility or department to both effective (jobs are done right) and efficient outputs (profits are earned) that meet the company's objectives.

A manager who uses planning, organizing, controlling, and leading properly will be in command of operations. This will allow the manager to be in a position to take care of the unexpected events that occur (such as the customer's automobile is wrecked by a technician when on a test drive and various details such as insurance must be handled). Although uncommon events will not always have a procedure, recent history has made emergency procedures a necessary part of a service facility's operation manual. Specifically, how will management duties be shifted so the organizational structure can be maintained? Who will do what and when? Specifically, emergency procedures must identify employees who would have some level of authority to take action within the reasonable parameters. Certain employees also should be given other assignments (such as someone to call 911, someone to collect information for the emergency service personnel and insurance company, someone to take care or check on customers, someone to call the owners or attorney, and so on). In the case of a severe catastrophe (such as a fire or robbery), again employees should know what to do without being told. In fact, these emergency procedures should be practiced on a slow business day, and incorporated into new-employee induction programs as well as employee training.

Concluding Thoughts

This chapter concludes the first part of the book, which has introduced the basic organizational and business characteristics that define automobile service facilities and the jobs of service managers. Of course, many books have been written on each of these topics and people who are or aspire to be service managers are encouraged to examine each more thoroughly. Knowing how to work with and within the organizational and business environment is important not only to the performance of the service manager but also to the success of a service facility. Therefore, the ability of service managers to work with these organizational and business features will typically have an impact on their performances and the corresponding evaluation of their performance.

As noted in this chapter, operational procedures and the operation manuals should describe of the work to be conducted in a facility or service unit. This information offers service managers expectations and standards, which can be used to evaluate the performance of the facility or unit and the people in it. If a service manager has a concern or does not believe the outcomes are acceptable, the manager must dig deeper into the work being done by the employees. One check is to examine the input, process, and output stages of the operational system.

In addition to the review of the system, a service manager may need specific information about what an employee or employees are supposed to do. If an employee is suspected of performing poorly, are they actually doing poor work or are they doing work they should not be doing? For this reason, job descriptions are required. Possibly, a low-tech employee is trying to do work that is beyond his or her capability without additional training and practice under an experienced technician. It is even possible that a high-tech employee is not doing high-tech work and feels that the routine low-tech work is not worth the effort. Again, a job description is needed.

Consequently, after the operation manual is prepared around a system that will work for the service facility, job descriptions are needed for each classification of employee or every position at the service facility. This would include, at most facilities, the service manager's position, the service consultant position, the technician positions, the parts specialist position, and the cashier position. The descriptions are needed so the service manager can properly recruit, hire, induct, develop, and evaluate people to become employees within the system that was set up.

At large facilities, job descriptions are prepared by a human resources department, but at smaller facilities the service manager may be expected to do them. As explained in the next chapter, the preparation of the descriptions is not a complex or difficult task, but these descriptions must be written and contain the detail necessary to be useful. This means that specific job tasks to be performed by each position should be outlined so that the service manager can exercise control and the operation of service facility will coincide with the operation manual.

Review Questions

1. For a manager, explain the purpose of:
   A. company policy
   B. operational procedures
   C. an operation manual
2. Explain why company policies and the operational plan are connected to operational procedures and operation manuals.
3. Explain why a manager would want to create a formatted system.
4. Describe each stage of a business system and how feedback is used by management. Why is it important?

PART I

PRACTICAL EXERCISE

Small Group Breakout Exercises

You are the service manager at an automobile service facility/dealership. The business is an S Corporation. Your services include general diagnostic work, repairs, and maintenance on all models.

You report to the two owners of the corporation. One owner is the Corporate Executive Officer (CEO) and also performs the duties of a general manager in charge of service operations. The other owner is the Chief Financial Officer (CFO) and also performs the duties of the business manager. They share an administrative secretary.

As the service manager, you have two people reporting directly to you. The positions of the two people are: Service Consultant and Parts Specialist (at this dealership, the CEO has required that the parts specialist report to you even though his office is in the parts department). In addition, there are five technicians employed in your department: one A tech, two B techs, and two C techs.

The facility contains five fully-equipped service bays that have a workbench, vise, lift, compressed air hose, and access to water. In addition, there is a room devoted to the storage of other shop equipment such as jacks, stands, and special tools. There is also an oversized work area that has tire-mounting and balancing equipment, a brake lathe, and a solvent tank that sits next to a service bay in which there is an alignment machine.

Management feels that the dealership facility can provide a customer with a wide range of services. In addition, they feel that they meet the automobile manufacturer's franchise agreement requirements with the exception of towing and major body repairs, which must be subcontracted to outside repair facilities.

To assist the CFO, there is a bookkeeper who supervises one full-time and one part-time cashier. In addition, the dealership employs a maintenance supervisor who handles facility repairs as well as two part-time employees responsible for keeping the building clean and the grounds maintained.

The owners have determined the business needs more capital to expand in the future and, as a result, must prepare a new business plan for the bank. They requested that you, as the service manager, provide the following information that will be included in their plan:

1. An organizational diagram for the service facility with a description of the chain of command for all line and staff personnel.
2. The job title, basic duties, and basic qualifications for everyone who reports to you—specifically each level of technician (A,B,C), the service consultant, and the parts specialist.
3. A description of the formal lines of communication procedures to be followed in your department. Support the rationale for your procedures with citations from the appropriate textbook chapters.
4. Your suggestions on appropriate mission, goals, and objective statements for the facility. To assist you in preparing your response, the CEO stated that he wants to see the dealership increase sales in all departments by 10 percent a year over the next five years. Support the rationale for your suggestions with citations from the appropriate textbook chapters.
5. For the service department, the CEO needs you to prepare a set of goals and objectives with targets and benchmarks for both the service and parts sales operations. To assist you with this question, currently, the service department has sales of service and parts of approximately $75,000 per month and each technician exceeds their workload capacity. In terms of the manufacturer's customer satisfaction score for your dealership, you earned a rating of 80 percent. The most common complaint about your service department is the amount of time it takes to get an appointment to have a warranty repair fixed.
6. Draw a diagram that describes the work flow system in your department from the time a customer calls for an appointment or comes in to purchase a part to the payment of the invoice. The diagrams may include special or important features, and must be in support of the goals and objectives you prepared for Question 4.

PART II

PERSONNEL

CHAPTER 5

HUMAN RESOURCES MANAGEMENT

LEARNING OBJECTIVES

Upon reading this chapter, students should be able to:

- *Explain the purpose of a job description.*
- *Prepare a job description.*
- *Explain the difference between staff exempt and nonexempt employees as per the Fair Labor Standards Act.*
- *Describe the differences in the task analyses for an A, B, C, and D auto technician.*
- *Outline the primary tasks in the personnel management process and explain the purpose of each.*

Introduction

In the strategic plan, an operation manual explains what each work unit at a service facility will do to meet the plan's objectives. The manual explains to employees what their unit will accomplish and how it will do it. In order to achieve the expectations of the unit, each job in the unit must have a description explaining what is to be performed by the employee in that position. The job descriptions must go together like the parts of a machine and run smoothly in order for the work unit to function effectively and efficiently.

Therefore, the success of a work unit depends on employees knowing what they are to do. Specifically, people cannot be successful if they do not have this information, and work units cannot be successful if the duties to be performed are not in anyone's job description. For this reason, as shown in Figure 5-1, job descriptions are at the base of the strategic plan's pyramid. In order for a manager to plan, organize, lead, and control, he/she must have descriptions for each position in the work unit. These descriptions must include all of the work tasks to be conducted in the unit. In this case, the unit can mean that the service department is structured as a group or team (technicians and a service consultant), as a dispatcher system, or as a manager who also serves as a service consultant and provides daily oversight. The primary purpose of this chapter is to explain what must be included in a job description, how it is prepared, and why it is critical to personnel management tasks.

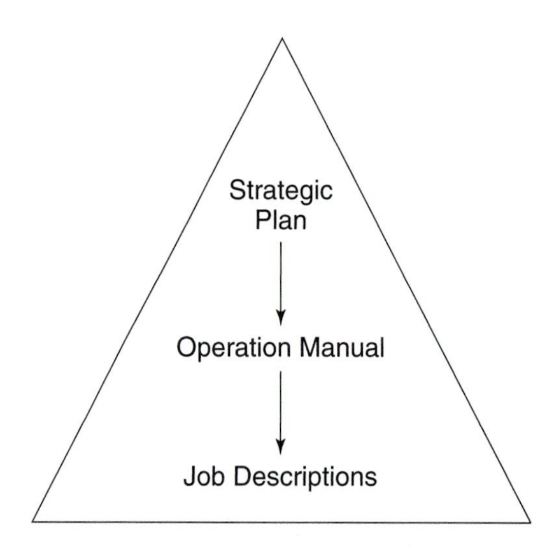

FIGURE 5-1 Job descriptions: base of the pyramid.

Job Description

A **job description** presents the work to be conducted by a person who is to be hired to fill a specific position. As a result, a job description must identify the following:

- The work unit and title of the position
- The classification or rank of the position
- Qualifications required, including:
  - education
  - certification and licensing
  - work experience
- Job tasks and duties to be performed.

Job Titles

A job title is important to a person looking for a job and to the person who holds it. For instance, when someone looks at a job advertisement, they first notice the job title. When people talk about what they do at work, they usually use their job title. In addition, when someone needs information from a work unit, they will likely use the name of the position when seeking it. For instance, when managers or owners have a question about an invoice, they may ask to talk to a bookkeeper or cashier. If there is a question about a repair order, they may ask for the service consultant. In other words, job titles indicate what a person knows or is likely to know about something.

Therefore, the title given to a job has many uses. It must accurately indicate what a person in a particular position does. A title should not misrepresent what a person does in an attempt to make it sound lofty, such as a professional automotive analyst instead of the popular title automotive technician. When this occurs, it may cause confusion and discourage good people from applying for a job because the job title does not make sense. It may even generate jokes about it.

When a job title is a common title, it should indicate the nature of the work to be conducted in a position. For instance, the title technician is no longer appropriate as a stand-alone title because there are so many types of technicians. Consequently, a title must include descriptive terms, such as auto technician, computer technician, electronic technician, and so on. In some cases, a title may need additional descriptors to accurately portray the position, such as automotive transmission technician or collision repair technician.

Usually, the next item of importance is the department or work unit in which the job is located. This gives insight into what a person in a position does. For instance, an automotive technician may be assigned to the service department or could be assigned to the collision repair department to perform mechanical repairs (engine, electrical, brake system, etc.) on wrecked automobiles. Both positions may have the same **job duties** and pay, and therefore the company will prefer to use the same job title. So the name of the work unit is critical to the job description. Therefore, a title may be modified to include the work unit, such as Automotive Technician: Collision Repair Department.

Job Classifications

When appropriate, a job title also may include additional classification information that may constitute a rank or standing that is relevant to the position. For example, the title of Lead Automotive Technician would indicate the employee that serve not only as an automotive technician but also in the first position for some level or responsibility when compared to other automotive technicians in the group. **Job classification** information included in a job title also may represent proficiency levels. In some cases, an employer will base the job title with proficiency information on a specific business need. For example, a transmission shop may have a need for an import-vehicle transmission rebuilder and a second one for domestic-vehicle transmission rebuilder. Each would be proficient in the type of work the service facility needs completed. Other job titles that indicate proficiency level are common across all automobile businesses. With respect to automotive technicians, a common classification is to designate four levels: *A techs, B techs, C techs* and *D techs*. The differences between these grade levels indicate the jobholder's proficiencies. Typically, it would be up to an employer or labor union contract to precisely determine what qualifications a person would have to have to be placed in one of the four categories. As a general rule, these are probably the most common qualifications expected by many employers:

- An A-tech position would require the employee to repair, maintain, and diagnose (to the highest levels of Bloom taxonomy: analysis, synthesis, and evaluation) all systems of an automobile. Typically, some type of formal education (often taken at the postsecondary level, from a NATEF master certified school or manufacturers college program such as Ford ASSET, Toyota T-Ten, General Motors ASEP, among others) in automobile repair is expected, as well as enough relevant work experience. A technicians typically hold ASE Master Technician Certification (perhaps even additional certifications such as L1 and L2 when appropriate), as well as any state-required licenses. Depending on the employer, manufacturer training and formal certificates (or training courses) are required to fulfill warranty repair requirements. Often these technicians choose to specialize in a certain type of automobile repair even though they can still perform repairs on all systems of the automobile. In terms of the types of automobiles worked on, they again will typically be most comfortable with certain manufacturer makes (even certain models) but are capable of working on all makes and models, although some research and training may be required. In general, these technicians are the most analytical of all of the levels and are relatively rare to find. Although businesses like to say they have an A technician, most actually have a very good B.

- A B-tech position would require the employee to repair, maintain, and perform diagnosis on the systems of the automobile on which they have had training and work experience. These technicians have had some type of automotive training and may or may not have any formal education in automobile repair. Their work experience is typically extensive on the systems they have been trained to service, repair, and diagnose. Some B technicians are considered specialists on certain systems or manufacturer makes and are as competent (or even more competent) to make the diagnosis and repairs as an A technician. B technicians typically hold ASE certification in the systems in which they have expertise, which can be as few as three areas to as many as five or six. It also is not unusual for a very good B technician to be well-rounded in all systems of the automobile and even hold all eight ASE certifications (master technician), although they really perform extensive repairs and diagnosis in the ones they are most comfortable. B technicians also will hold all state-required licenses and, depending on the employer, manufacturer certificates (or training courses) are held to fulfill warranty repair requirements for the systems they repair. B technicians are typically very competent and efficient in the systems and automobiles they have experience repairing. They are more common than A technicians and most service facilities employ at least one B technician.
- A C-tech position would require the employee to repair and maintain a limited number of systems on an automobile. They can typically perform diagnosis on specific system components and have some ability to perform limited diagnoses on the system itself. C technicians will usually have some training and maybe even some formal education in automobile repair. They may or may not have much work experience. If they do have extensive work experience, it is often limited to a certain type of repair repeated over and over again. Depending on the business, a C technician may be considered a specialist on at least one system. However, if a repair becomes too complex because of unexpected problems, an A or B technician may need to provide assistance. C technicians may hold ASE certification in the systems in which they have expertise, which can be as few as one area to as many as four. It is not common for a C technician to hold ASE certification in all eight areas. C technicians may hold state-required licenses and, depending on the employer, manufacturer certificates (or training courses) may be held to fulfill warranty repair requirements for the systems they repair. C technicians who are experienced are typically competent and efficient in the systems and automobiles they have experience repairing. They are the most common technician working at a service facility.

- A D-tech position would require the employee to maintain a limited number of systems on an automobile. They typically can perform light or simpler repairs on specific components and have some ability to perform limited diagnoses on components. D technicians may have some training and maybe even some formal education in automobile repair. They may or may not have much work experience. When performing a repair, typically the oversight or assistance of a B or A technician is required. In some cases, D technicians may be considered a technicians helper or technician in training because they assist a B or A technician during a repair process. D technicians do not typically hold ASE certification, although they may hold or have passed as few as one area to as many as three or four (if passed, they typically lack the two years of required work experience to become ASE-certified in the area). D technicians may or may not hold state-required licenses and typically do not hold any manufacturer certificates (although they may have taken some basic manufacturer training courses). D technicians who are experienced are typically competent and efficient in the maintenance and repairs of the systems they have experience servicing. They are common entry-level employees and technicians who specialize in maintenance and light repairs.

Positions that require greater proficiency usually command significantly higher wages or salaries. For instance, an employer may classify their secretaries in terms of numbers. A grade 1 secretary has the least education, training, and experience and has the lowest salary, whereas a grade 5 has the highest salary because he/she has the most education, training, and experience.

Job Rank

The **job rank** signifies the authority or responsibility assigned to a position. For example, titles such as director, manager, coordinator, and supervisor stand for some level of authority that would be assigned to the job. Job ranks also may suggest some form of status within a job group, such as senior service consultant or lead technician. Although these titles do not necessarily imply that any authority is assigned to the position, they do suggest that some level of leadership, experience, and expertise would be required for the person to occupy the position. In cases in which a rank is included in a title, it is implied that additional money is paid to the person who holds it.

Pay, title, and responsibilities are used to determine whether a job is **staff exempt** or **nonexempt**, which are classifications given by the federal government in the *Fair Labor Standards Act (FLSA)*. Although this text cannot cover the details of the FLSA, the basic intent of the act is to place all workers in the United States into one of these two categories: those who are not exempt (nonexempt) from the regulations and those who are exempt. The most important difference is that a nonexempt employee must be paid at least the federal minimum wage (possibly the state minimum wage if it is higher) and overtime (one and a half times the employee's hourly wage) when the employee works over 40 hours per week. Staff exempt employees do not have to be paid overtime; however, the exempt employee must be paid at least the minimum amount per week specified in the current law, which is $455 per week under the current law. Usually nonexempt staff are paid hourly, while exempt are on salaries. If a nonexempt employee is on a salary, the hourly rate must be calculated and the employee paid one and a half that amount as overtime.

Managers with personnel responsibilities *MUST* understand the FLSA, which began in 1938 and has been amended several times since then. The last time the job requirements were changed for staff exempt and nonexempt employees was in 1949 and the minimum amount to be earned by staff exempt employees was updated in 1975. Currently, employers are required to comply with new FLSA regulations that went into effect on August 23, 2004. According to the law, exemptions would not apply to manual laborers or other blue-collar workers who perform work involving repetitive operations with their hands, physical skill, and energy. They also do not apply to nonmanagement production-line employees and nonmanagement employees in maintenance, construction, and similar occupations such as carpenters, electricians, mechanics, plumbers, iron workers, craftsmen, operating engineers, longshoremen, construction workers, and laborers.

The act does not limit the number of hours in a day, or days in a week, an employee may be required to work, including overtime hours, if the employee is at least 16 years old. In addition, the act does not state any requirements on vacations, holidays, severance or sick pay, meal or rest periods, premium pay for weekend or holiday work, pay raises, fringe benefits, discharge notices, reason to be given for discharge, or immediate payment of final wages to terminated employees. However, some states have laws that cover some of these issues, such as meal or rest periods and discharge notices. Employers must comply with their state laws. Furthermore, the act does not preclude employers from entering into collective bargaining agreements providing wages higher than the statutory minimum, a shorter workweek than the statutory maximum, or a higher overtime premium (double time, for example).

With respect to staff exempt employees, in order for a position to be exempt, its primary job duties must meet certain minimum tests and the person must be paid a salary that is more than the minimum amount (stated above at $455/week). Exempt employees, according to the act, are those employees in bona fide executive, administrative, and professional positions or those working as an outside salesman. The FLSA presents the job features that define each of these exempt categories (executive, etc.) as well as salaries. As a result, the FLSA requires that employers keep records on each position, such as job position, duties, pay, hours worked, and so on. In other words, an employer cannot give a title, such as administrator or a professional label, to a person who is in a nonexempt job in an attempt to get around overtime payments. The law does not permit it!

Job Tasks

Job responsibilities need to be broken down into the tasks to be performed by an employee. The tasks present the work that is to be conducted by the person in the position. The accuracy of the job descriptions is critical to the success of a facility. Coaches of sports teams know that players must know their assignments and then do them. This is the most important key to the success of the team and the reputation of the coach. The same is true for the service department and service manager.

A list of **job tasks** is usually prepared by a person who performs or has performed the job or by interviewing people who perform or have performed the job. For many years, vocational educators have used this task analysis approach to set up their curriculums and lesson plans. A task list is typically limited to major tasks to be performed in a job and usually consists of 10 to 12 tasks. For example, one task for an automotive technician may be to diagnose automobiles with operating problems. In this task, each type of different operating problems would not be separate tasks. If the technician's position were limited to alignments, the task would indicate that diagnostic work would be conducted on automobiles with alignment problems.

Examples of task lists for three job descriptions (used by the authors in the *Service Consultant: Working in an Automotive Repair Facility*) are shown in Figure 5-2 for a service writer, Figure 5-3 for a service manager, and Figure 5-4 for a parts specialist. Note that some of the tasks are to provide support for one of the other positions and that some tasks in one description will overlap the tasks in another position. This is to ensure that on very busy and hectic workdays employees could back each other up. This is called **cross-training** and means that employees can do each other's jobs. Therefore, if an employee was absent or had to leave the facility, another employee could pick up anything that needed to be done. For example, if the service writer/cashier was not available, one of the two other employees could receive a payment from a customer.

These task lists do not present job details, which are called duties. Typically, task duties are added under each task for teaching and training purposes. In addition, each task in a job description should be assigned a percentage of time that an employee takes per week performing the duties, such as an amount of time based on a 40-hour week. For example, if an employee spends four hours a week on a task, such as opening and closing the shop every day, the task would show that 10 percent of the employee's work time is spent on that task. An example of a list of duties and subduties for the first task in the service consultant's job tasks is shown in Figure 5-5.

Finally, a common question about job descriptions concerns the last task, which is always other duties assigned. Often this catch-all task is used when an employee is assigned a duty that is not on the job

Job Description
Service Consultant/Cashier

(date)

1) Opens shop at 7:30 a.m. and close at 5:00 p.m.
2) Greets customers, answers phone calls, provides information, makes appointments, calls customers for approval of work on vehicles, calls customers when jobs are completed, and places follow-up calls after repairs are made.
3) Presents invoices, reviews parts and labor charges, and receives payment from customers in the form of cash, check, or credit card.
4) Makes arrangements for customer shuttle.
5) Prepares and maintains Repair Orders (RO) and Computer Invoices (IN)
6) Prepares and monitors the Service Consultant's Progress Sheet.
7) Assists service manager to ensure service work is completed in a timely fashion.
8) Assists parts specialist as needed in preparing cost estimates, ordering parts, receiving parts orders, and checking parts invoices to ensure the charges are accurate.
9) Maintains inventory of office supplies including supplies for computers, photocopier, fax, credit card machine, printers, forms, reports, and other materials as needed.
10) Makes daily deposit and prepares accounting report.
11) Closes building at the end of the day and checks cars in parking lot.
12) Other duties assigned.

FIGURE 5-2 Service writer who also serves as a cashier task list.

description. While this is the appropriate use, this task should not take up more than 5 percent of an employee's time. In fact, this task should amount to less than 5 percent. If an assigned *other duty* is taking more than 5 percent of the employee's time, then it should become a separate task and the percentages allocated to other tasks should be adjusted. Keeping the percentages of time allocated to each task in balance is important for two reasons. First, if the percentage does not come up to filling a 40-hour week, then possibly the job should be a part-time job. Second, keeping the percentages in line is one way to ensure that nonexempt employees do not have to work overtime on a regular basis.

Job Description Tasks
Service Manager

(date)

1) Opens up shop area at 7:30 a.m. and close at 5:00 p.m.
2) When appropriate, test-drive vehicles for diagnostic purposes and postservice checks.
3) Reviews work orders and distributes work.
4) Assign service orders and minor repair work to "service" technicians and monitor their work for quality.
5) Prepare cost estimates for parts and labor for repairs and record on job estimate forms.
6) Orders and receives parts.
7) Ensures part invoice charges are accurate.
8) Returns unused parts and cores and ensures proper credits are awarded.
9) Assists Service Consultant in keeping the Progress Status Sheet up-to-date.
10) Reviews inventory of supplies (including oil filters), oil, parts, and other materials and places orders as needed.
11) Ensures that service technicians keep work area clean and empty shop garbage cans.
12) Makes sure that all customer vehicles and shop shuttle are secured at the end of day.
13) Recruitment, selection, induction, development, and appraisal of supervised personnel.
14) Other duties assigned.

FIGURE 5-3 Service manager who also serves as the parts specialist task list.

Personnel Management Tasks

The job description is a key element in personnel management. Personnel management responsibilities have traditionally been divided into segments that are assigned management tasks. These tasks are: **recruitment**, **selection**, **induction**, **development**, and **appraisal**. As shown in Figure 5-6, one task leads to another except for the last two (development and appraisal), which keep recycling. The thread that runs from the first task (recruitment) to the last (appraisal) is the job description. Without job descriptions, the tasks cannot be performed and, if they are, any actions that are taken cannot be valid.

Job Position Description Tasks

Parts Specialist

1) Prepare cost estimates for parts and labor for repairs and record on job estimate forms.
2) Check estimated costs with technicians.
3) Order parts.
4) Receive parts and checks deliveries with orders to ensure parts are correct and invoice charges are accurate.
5) Enters part numbers, names of parts, and costs plus labor charges on the front of job ROs.
6) Makes photocopies of delivery invoices.
7) Delivers parts to production manager or technician.
8) Assists Service Consultant in keeping Status Sheet up-to-date.
9) Returns unused parts and cores and ensures proper credits are awarded.
10) Keeps working area clean—includes dusting all areas twice a week, vacuum and wipe clean the parts counter once a week, and vacuum floor once a week.
11) Other duties assigned.

FIGURE 5-4 Parts specialist task list.

When personnel management is assigned to a manager, the tasks are the job description tasks for that manager. The responsibility of the manager, therefore, could be to:

- Prepare a plan for the duties for each job description task to be performed.
- Organize duties and sub-duties to accomplish each task.
- Control duty and sub-duty performances each.
- Lead people in the unit through the plan.

A brief overview of each task follows. The details for the tasks are presented in the remaining chapters in this part of the book.

Recruitment

Using the job description, the objective of the recruitment task is to attract qualified people to apply for a position. Before any candidates can be recruited, however, managers must have the title and up-to-date job requirements (list of tasks) for the position plus the qualifications (education, work experience, certificates, and licenses) to be met by applicants. The title, job requirements, and qualifications are used to write a brief and precise description to be sent to advertisers and employment agencies.

SERVICE MANAGER – DUTIES AND SUB-DUTIES

TASK #1: OPENING RENRAG AUTO REPAIR

1. OPEN BUILDING, FRONT OFFICE, & WAITING ROOM
   1. Enter front side door, turn on lights, and turn off security system.
   2. Turn CLOSED/OPEN sign around to show open.
   3. Turn on lighted OPEN sign.
   4. Unlock front door.
   5. Turn on computer and printer.
   6. Turn on TV security monitor.
   7. Switch answering machine to position A.
   8. Turn on photocopy machine.
   9. Turn on radio.
   10. IF WINTER, turn thermostat up to 72 degrees.
   11. Enter alignment bay.
2. OPEN ALIGNMENT BAY
   1. Turn on lights.
   2. If WINTER, turn thermostat up to 65 degrees.
   3. Turn on light in bathroom.
   4. Go to inside bay.
3. OPEN INSIDE BAY
   1. Turn on bay lights and lights to storeroom.
   2. Enter compressor room.
4. OPEN COMPRESSOR ROOM
   1. Turn on compressor.
   2. Enter back bay.
5. OPEN BACK BAY
   1. Turn on back bay lights.
   2. IN SUMMER, open bay door on the left (not facing Arch Street).
   3. IN WINTER, turn thermostat up to 65 degrees.
   4. Enter lube bay.
6. OPEN LUBE BAY
   1. Turn on bay lights.
   2. IN SUMMER, open back bay door.
   3. IN WINTER, turn thermostat up to 70 degrees.
   4. Check for key drop envelopes.
   5. Enter front office area.
7. OPEN INTERIOR OFFICE
   1. Turn on lights.
   2. Uncover computers.
   3. Turn on computers and printers.
8. FRONT OFFICE CHECKS
   1. Listen to phone messages, check parking lot, and proceed to write up repair orders.

FIGURE 5-5 Service manager duties for task one of the service manager task list from Figure 5-3.

9. PARKING LOT - CUSTOMER VEHICLE INVENTORY
   1. Use the customer automobile inventory sheet to check all customer cars left overnight in the building or in the lot.
      Check customer automobiles left overnight for damages and missing parts not noted on the customer automobile inventory sheet.

FIGURE 5-5 *continued*

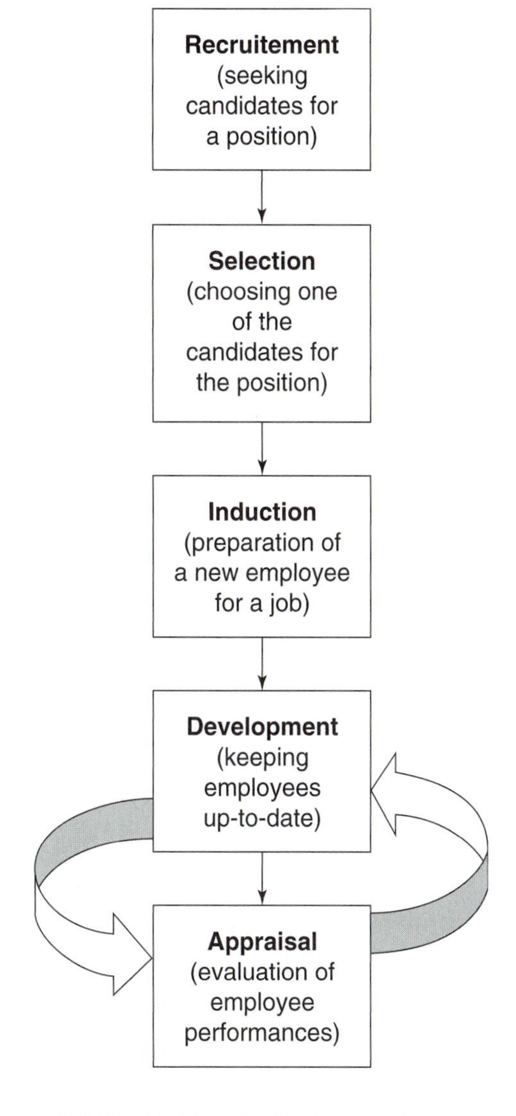

FIGURE 5-6 Tasks in the personnel process.

Selection

Using the job requirements and qualifications for a position, the selection task's purpose is to take the applicants and hire the best-qualified candidate for the job. To identify the best-qualified candidate, managers must use the job requirements and qualifications to conduct a paper review of the applications and pick the people to be interviewed. Before meeting with the candidates, the manager should use the job requirements and qualifications to prepare the questions for the applicants, all of whom should be asked the same questions. Next, the interviewer or interview committee must be selected and the method to be used to interview the candidates must be chosen. Notes on the interview of each candidate must be taken and kept for a period of time. In addition, in the selection stage the manager must know the pay range for the position, and how the hiring decision will be made. After a person is selected to be hired for the position, managers must then know who will make the job offer and how it is to be made.

Induction

Using notes from the interview, the manager must set up an induction program for the new employee. First, the induction of a new employee is to help the person get accustomed to his/her new position and become familiar with the way the company conducts operations. Second, using the information gained in the interview, an induction plan should focus on the new employee's ability to perform all of the tasks listed for the job. For instance, if a new employee appeared to be weak performing a job task, the manager should prepare a plan that will assist the person to learn how to do the work. Because an induction period includes a learning period, which is a period of time when the new person must demonstrate he/she is capable of performing all of the tasks responsibly, the new employee will either be awarded a permanent position (with a wage or salary increase) or terminated.

Development

The primary focus of employee development is to keep employees up-to-date on the latest developments related to the tasks and duties required in their job, such as technology, new certification expectations, license changes, revised or new inspections regulations, and so on. In addition, development programs should keep all employees up-to-date on changes in the operation manuals and company procedures and policies. Any substantial commitment to an employee (such as time off and travel reimbursement to attend a workshop) usually occurs after a new employee has completed an induction program. At the same time, however, a new employee would participate in programs concerning changes in company manuals or policies plus any workshops given at the facility (such as the use of a new machine by a manufacturer). Therefore, managers should create development plans for each employee (based on the person's job description), the facility, and the work unit. When preparing a plan or before it is to be put into effect, managers should seek employee input and reviews. In all cases, development programs must be monitored and evaluated.

Appraisal

Appraisal of employee performances must be ongoing (not an annual event) and in reference to the tasks in a person's job description. Without the job description, the supervision of performances will likely vary from day-to-day and from one supervisor to another. In other words, supervisors must know what to supervise (not what they like and do not like). Appraisals should be **formative** and **summative.** Formative appraisals are expected to improve performance. Records of these appraisals are not kept since they require an employee to honestly seek and be willing to receive assistance in a nonthreatening relationship with a supervisor. A summative appraisal, however, is a formal evaluation used in promotion decisions, pay increases, or terminations. Records of summative reviews must be kept and discussions with employees must be recorded, as shown in Figure 5-7.

FIGURE 5-7 Records of formative reviews must be kept and discussions with employees must be recorded.

Collective Bargaining Contracts

If employees are covered by a **collective bargaining** (union) contract with the employer, the personnel tasks may be affected. For example, job openings may have to be posted internally for all employees to see before they can be announced publicly. In addition, there may be some restrictions on the hours worked, lunch breaks, vacations, sick time, and so on for nonexempt employees. In other words, even though the federal or state government may not require limits or benefits for nonexempt employees, the union contract might. As a result, managers must read the current union contract and be sure to work within its regulations, especially the grievance process. A technician's copy of a Teamster Union handbook (labor agreement contract) for a New Jersey Car Dealer is shown in Figure 5-8.

The Three Most Important Management Responsibilities

Before entering into a full description of the tasks to be performed in personnel management, the three most important responsibilities of a manager must be listed. When the manager conducts the recruitment and selection tasks, these responsibilities are especially important to the induction and development tasks. These three important responsibilities are to:

1. Document
2. Document
3. Document

Without proper documentation, the induction program does not have a starting point. The credentials and interview notes on a new employee are the basis for: (1) an induction program that will assist the person in the performance of his/her job to the satisfaction of the employer and (2) for a development plan to keep the company and employees current. Specifically, assume that a person interviewed for an alignment technician job; however, he did not have any experience working on the alignment equipment the service facility owns. The manager should realize in the interview that in order to perform the job to the desired level of performance, an induction program would be necessary. This may require training, guided practice, or shadowing under a technician who knows how to use the equipment until the technician's efficiency can be improved.

Therefore, documentation is important to a supervisor when hiring or when a problem occurs, as it is the only protection a manager has when he/she must take personnel or financial action. For example, when a person complains that their application was not reviewed, an employee has to be terminated, or a customer complains that work was not

FIGURE 5-8 Teamsters' Union handbook.

performed as requested, the only defense is for the service manager to produce documents. The documents must show that appropriate action or work was or was not performed according to directives set forth in company policy, state law, or industry standards. If a service manager or owners must go before a labor negotiator or mediator in response to a dispute over a personnel action, the only defense the negotiator is interested in is documented evidence.

Therefore, a manager who wishes to act professionally should have copies of applications and employee credentials, resumes, references, notes on interviews and meetings (with dates), and copies of memos and e-mails regarding directives. The manager should keep a personal diary of observations with dates and details about what happened, what was said, who was involved, and what actions were taken. Also, in response to agreements or disagreements in a personal conversation or over the phone, the manager may need to write a follow-up memo or e-mail restating the contents of the conversation. Time after time, managers learn that some people's memories (intentionally or unintentionally) are not strong enough to keep the details straight and the only way to ensure the facts are correct is to have proper documentation. This is especially true for understandings about induction and development agreements and performance expectations.

Final Thoughts

The tasks to be performed in the management of personnel operations are dependent on the job descriptions. Those tasks assigned to the service manager become tasks in the job description for the service manager. As a result, the tasks and their duties, and subduties, must be taken seriously. They cannot be ignored or made up when they are to be conducted. The success of a manager depends greatly on the success of the people who work under the supervision of the manager. If the subordinate workers do not do their job properly, this reflects directly on the perception of the manager's competence. Sadly, however, if subordinates do not do their job as expected, the manager is very likely incompetent!

Therefore, the duties and subduties for each of the personnel tasks are important to the business and to each employee. Of course, managers may have preferences (as the authors do) for conducting these assignments in each task. Regardless of the methods followed to meet the objectives of a task, managers have the responsibility to make sure that the company and employees are properly represented and not harmed.

Consequently, when managing the personnel process, managers should make business decisions and never personal or emotional decisions. For example, appraisals are to determine if employees are able to perform their job tasks. Personal opinions or emotions cannot enter into a decision about whether a person can or cannot do a job. The decisions and actions of a manager must be business decisions. Liking or not liking a person has no place in a manager's life. Of course, managers are human and humans become irritated and angry. When this occurs, managers must walk away from the situation and make a decision or take action at a later time. When management personnel decisions are not based on whether a person is doing the job assigned by a company or are made out of frustration, the company loses, a person (employee) will likely be unfairly harmed, and the manager has not done his/her job.

Review Questions

1. Explain the purpose of a job description.
2. How is a job description created?
3. Explain the difference between exempt and nonexempt employees as per the Fair Labor Standards Act.
4. Describe the differences that might be found in a task analysis for an A, B, C, and D auto technician.
5. Outline the primary tasks in the personnel management process and explain the purpose of each.

CHAPTER 6

RECRUITMENT AND SELECTION OF NEW EMPLOYEES

LEARNING OBJECTIVES

Upon reading this chapter, students should be able to:

- *State the objective of the recruitment task.*
- *List and describe the duties to be performed in the recruitment task.*
- *State the objective of the selection task.*
- *Explain what the term best qualified means and why it is important to recruitment and selection.*
- *List and describe the major duties to be performed in the selection task.*
- *Explain the difference between the standard and open-ended interview.*
- *Outline a recruitment and selection plan for a technician in an automobile service facility.*

Introduction

The personnel tasks are often assigned to a service manager's job description. If a task is not assigned to the manager, he/she often must carry out the task or many of the task duties anyway because he or she is in charge. As a result, a service manager needs to know all of the duties assigned to each of the personnel tasks. The purpose of the following three chapters is to describe the duties and subduties in each personnel task. The purpose for this chapter is to present the duties and subduties of the recruitment and selection tasks. For a manager to recruit and select personnel, all of the duties must be performed. The subduties have some options, which are discussed below. The order in which the duties and subduties are performed can vary from facility to facility. While this often depends on the preferences of the owners or upper management, the concern is that they are all conducted properly.

The recruitment of employees requires a job title, job description, and specific qualifications for the position, as discussed in the last chapter. Essentially, this information about a position is the blueprint for the job. Needless to say, a manager cannot recruit or select the best person for the position without the blueprint. When no one fits the blueprint, the recruitment process must begin again. After a person is identified and hired, the comparison of the person to the blueprint is used to prepare what is known as an induction and development plan for the new employee, as discussed in the next chapter.

Recruitment

The specific job requirements and qualifications for a position must be available before the recruitment of a candidate for a position can begin. For example, when a technician is needed, the specific duties he or she must perform must be identified and the qualification to do the job must be known. This information is required to meet the recruitment task's primary objective, which is to find qualified applicants. There are four basic duties to be performed in the recruitment of new employees:

1. Write the job announcement.
2. Advertise the position.
3. Collect applications.
4. Prepare application materials for screening.

Writing the Job Announcement

The job announcement must be written carefully to ensure that the proper information is presented to potential applicants. It also must be brief because it will be placed in advertisements in which space is restricted (and expensive) and, also, because it must gain the attention of potential applicants. Job announcements must include at least three pieces of information, including:

1. Position title and responsibilities
2. Specific job qualifications
3. Work experience

The job announcement also may indicate if references are or will be needed. In some cases, a facility may include its name and phone number, while others may decide to use a post office box. The reason for this decision varies; for example, the owners may plan to fire an employee once a replacement is identified or the facility may not wish to inform their competition about a vacancy in the facility.

An important feature about a job announcement is that it should attract qualified applicants and not people who are just looking for any job. A properly prepared announcement with clear minimum requirements can save managers considerable time if it sorts out job seekers who are not qualified. In addition, a poorly prepared job announcement may attract people who are overqualified. Although overqualified people are exciting to interview and can definitely do the job, they typically expect greater wages or salary than the service facility can pay. Even when they take the job for what a service facility can pay, they typically leave once they find a better job, and the entire time-consuming, and sometimes expensive, process begins again.

A job announcement should permit employees at the facility to repeat the same information over the phone to prospective candidates. In one situation, an improperly prepared job announcement caused a flurry of calls to a service facility about a job opening for a service consultant's job to be held in an unprofessional manner. The person who answered the phone was not qualified to answer any questions about the job but did so anyway. Because incorrect information was given to the callers, people who were not qualified for the position submitted applications, had references sent in, and made follow-up calls to the owners requesting an interview.

Therefore, a good job announcement begins with the title and description of the position and lets potential applicants know exactly what the person is expected to do on the job. For example, an announcement for an automobile technician needs to indicate if the technician will have to make general or specialty repairs (such as brakes). In the case of a service consultant, an announcement should indicate if the person will work with a team of technicians, what type of work is conducted by the facility, and if service sales are a big part of the job.

Next, the announcement should indicate the level of expertise, certifications, licenses, training, and education expected. For example, in the case of an opening for a technician, the announcement should indicate if the position is for an A, B, C, or D technician, and if ASE certifications and inspection licenses are required. This information will immediately attract some applicants and eliminate others.

Finally, the announcement must give some idea about the amount of work experience needed for the position. If the position is an entry-level position, this information will hopefully discourage applicants who are simply looking for a position that will offer higher pay. At the same time, if a person with special experience is being sought, then the announcement should state that fact so that only qualified applicants will apply.

As a result, the announcement must focus on seeking *qualified* applicants. In the previous chapter, a reference was made to hiring the *best qualified* applicant. Recognizing this difference is critical. The objective of the recruitment task is to attract qualified applicants, while the selection task must focus on hiring the best of the qualified applicants. If this is not clearly the intention of these two tasks, then a winning team that will make money for the facility will not be put together. In other words, if none of the applicants are qualified, then no one should be interviewed. The announcements must be sent out again because it is cheaper to rerun the search than it is to hire the wrong person.

An example of a poor advertisement is:

> Auto technician wanted. Call XXX-XXXX between the hours of 8 and 5.

The lack of information in this announcement tells potential applicants a lot. It says the owner or manager does not know what they want, or does not care, or does not know much about the work conducted by technicians. It also says that the owner/manager would not take the time to write a proper advertisement that would avoid wasting an applicant's time. This, therefore, suggests the owner/manager may not care much about the treatment or inconvenience caused to other people, including their employees. A more appropriate announcement would be:

> Auto technician; B-level to do state safety inspections and related repairs. Current ASE certifications in Brakes, Suspension, and Steering required. State safety and emission inspection licenses. Five years of full/part-time work experience. References required. Flat-rate pay system with a guarantee, retirement plan, health care, and uniforms provided. Call Mr. Joe, Manager, at XXX-XXXX to set up an appointment to fill out an application.

This advertisement will help to eliminate people who are not qualified, such as an entry-level technician or someone without any certifications or licenses. At the same time, it suggests the facility has some interest in an employee's well-being by offering benefits. It also encourages applicants to contact a specific person at the service facility at which time additional information can be provided, as seen in Figure 6-1. When the contact person is not available to take the phone calls from interested technicians, then the person answering the phone should take the person's name and phone number. In some cases, the manager may delegate this duty to someone else.

FIGURE 6-1 Manager performing telephone screening of possible job candidates.

Regardless, every phone call must be screened to make sure that the technician meets the minimum qualifications for the position. Applicants should be given information on how to apply and questions should be answered about the position. The amount the person will be paid should not be told to anyone, especially over the phone, because it should be dependent on the person's qualifications. Rather, the facility should have a printed announcement to be read or given to people about the position and the facility, such as working conditions and the like. If a person making an inquiry over the phone or in person about a position sounds especially interested and qualified, the announcement may be given or sent to him or her. In the authors' experience, job advertisements and promotional advertisements often can generate calls from competing facilities in an attempt to get information about wages, salaries, and hourly rates charged to customers.

The Advertising Plan—Searching for Eagles

There is an old saying that "if you are looking for a new employee and you want to find an eagle, you have to search for one." In other words, a service manager wants to find a person who is smart, strong, and willing to work long, hard hours with minimum direction. Of course, eagles do not flock. Turkeys flock! Eagles have to be found. Too often managers throw out a general announcement, hope to find an eagle, and then pick the best turkey from the flock. Frequently, this person is minimally qualified and not able to meet expectations after being hired. In these cases, the manager will not be happy and the new employee will feel he/she is being harshly treated.

Therefore, the recruitment of new employees is referred to as a search. This is because strong employees must be found. They usually will not find you because many employers want them on their team. As a result, an advertising plan to announce a position must use multiple sources to get the information to all possible candidates about a job opening, such as the newspaper, public and private employment agencies, placement agencies, job centers, and teachers at vocational schools, plus personal contacts with colleagues and top people who hold a similar job.

In addition, all job openings should be posted internally for all employees to read for several reasons. First, all employees should be given the opportunity to apply for a job. An employee may not be qualified for the position, and the written announcement is a nice way to let them know about the qualifications without offending them. Second, employees also have good contacts and will likely pass the information on to people who are qualified. In the authors' experience, these informal contacts often provide an opportunity to speak confidentially with an eagle. Third, an employee who would be a strong candidate may be overlooked because he/she may be timid and does not seem like a strong candidate. These *sleepers* must be contacted and encouraged to apply.

When a trusted colleague who is knowledgeable about the field (such as the technical specialist, service consultant, and so on) refers an applicant (who is not related or a longtime friend), this is often a good endorsement. The manager must follow through and make contact with the person. Of course, the manager should not hire the person without following through on the remainder of the duties in the recruitment and selection process. In other words, the manager must make sure the person is qualified and avoid several risks. For example, the authors found that some personal referrals were because the person making the referral could not refuse to do it (the person could be a brother-in-law). In addition, when an applicant presents a glowing personal referral and strong credentials, care should be taken to determine that the applicant isn't a job jumper (person who moves from job to job, is never satisfied, and has held more than two or three different jobs in the past several years). Consequently, a personal referral must be reviewed by contacting former employers and people who know about the professional ability and activities of the applicant. A glowing referral, however, from another service facility must be regarded with suspicion. The former or current employer may be using it as a way to get an advantage over the competition by putting a poor employee in your shop or to get rid of a poor employee. As a result, while a personal referral may lead to an eagle, there are cases of being led to a turkey instead.

Collecting Applications

When job applications are received, a folder should be prepared for each applicant. Upon filling out an application, applicants should be given a checklist of the materials to be submitted. These materials are supporting documentation to verify the applicant's qualification. As materials are received, they should be placed in each folder. The checklist of materials to be submitted would include a résumé and photocopies of certificates, licenses, and diplomas (possibly even official school transcripts). In addition, if references are requested, this should be on the list as well. As the materials are received, they should be dated, time-stamped, and a checklist in the folder should be marked. In addition, the files of all applicants should be placed in a secure location for protection from people who are trying to get advance information, such as an employee who recommended a friend for the job. After all of the information requested from an applicant is received, then it is ready to be screened as part of the selection process.

Determining when a search has ended and selection begins is an important decision. In some cases, a search may be kept open until a sufficient number of qualified applicants have been received or when the owners or management believe they cannot delay any longer in filling the position. When a service facility that is part of a government unit (such as a fleet facility for a city or county) recruits new employees, they are to set a closing date when the opening is announced. This is because they are a public (not private) employer and are subject to federal and state directives.

Preparing Applications for Screening

At the end of the recruitment task, the applications are moved to the selection task. Technically, only the applicant files that have all of the required checklist materials should be transferred to the selection task. All incomplete files should not be considered or even retained. While this is the case in a formal search, it is not advised for profit companies for several reasons. The incomplete files may include some strong candidates who might need some extra time to complete their application. For example, the authors had one applicant who did not complete his file because his wife was to make the photocopies of his licenses and could not get to the store to make the copies because of a sick child. In another case, an applicant realized he did not really qualify for the position; however, the manager was impressed and his application was kept on file in case a position for which he was qualified became available. In addition, a facility with a good reputation will have people stopping in to inquire about a job. An application form should be given to these people and their credentials should be placed on file. These files should be reviewed and the ones that qualify for the open position should be transferred to the selection task.

When the files are prepared for the selection process or task, the issue is whether or not they are complete and, if incomplete, whether or not they should be used. At this point in the process, the temptation is for decisions to be made regarding the applicants qualification status for the position. Unless the reason for disqualifying a candidate is obvious, such as the failure to hold an inspection license, the candidates should be forwarded. There are two reasons for this practice. One is that the people collecting the materials from the applicants are clerical workers who often do not have the background or experience to make such decisions. The people involved in the selection process are qualified to make the decision because they understand the job that needs filled. Second, the review of the qualifications of the applicants takes *time*, which can slow down the processing of applications. This can create problems when time is of the essence. Filling a position is usually important to the production of the service facility.

Selection

Using the job description and qualifications prepared for the recruitment task, the objective of selection is to select the best qualified applicant. In the recruitment task, the emphasis is on the word *qualified*, while in the selection task the emphasis is on the word *best*. In other words, after all applications are reviewed, only those who are qualified are to be considered for selection. The manager or selection committee must decide which one is the best.

There are many techniques a manager may use to determine the best. For technicians, the manager may set up a hands-on test for applicants to demonstrate whether they can do the job. In the selection of a technician or service consultant who will work with a team of technicians, the team leader or other technicians may be asked to sit in on the interview and provide input. Regardless of how the best is determined, the four steps, or duties, to be conducted in the selection task are as follows:

1. Screening applications
2. Conducting interviews
3. Making a job offer
4. Follow-up

Screening Applications

The first duty to be performed in the recruitment task is to screen the applications. The purpose of the screening duty is to separate the qualified from the unqualified applicants. Note again, that unless an applicant is obviously not qualified, the identification of the unqualified candidates should occur in this task and not the recruitment task.

This is because the personnel who do the recruiting may not have not have the background or experience needed to examine the details in a résumé or on the application form. As a result, this should be left to the person or people who will select the best qualified candidate.

Therefore, the first subduty in the screening of applications is to eliminate those applicants who are not qualified. Next, the application folders must be reviewed to determine if all of the required materials are in the folder. If not, a decision must be made to either eliminate the applicant or to contact the person to see if the information can be made available. For example, in one case, an excellent candidate did not complete his application. Because he was such a strong applicant, he was contacted. He did not complete his application because he did not have a driver's license and he could not test-drive an automobile after conducting a repair or safety inspection. In addition, he could not be considered because the company's insurance carrier checks the driving record of all technicians regularly and new hires as well as candidates before they are hired. The loss of an employee's driver's license or a DUI conviction is a serious problem for a facility's liability insurance that covers company and customer automobiles (called a **garage keeper's policy**). In some cases, when a current employee cannot be covered by a policy, the owners must sign an agreement that the service facility will not let the employee drive a car. This is actually quite expensive as well as inconvenient for the facility. Candidates cannot withhold this information during an interview because insurance companies run a motor vehicle violation check on all new employees and an annual check of all employees when the insurance policy is renewed.

After the qualified candidates are determined, they must be ranked. The ranking should be based on a comparison of the candidate's qualifications to the job description. The ranking should not be a simple comparison but must take into account the work being done by the people at the facility. This is where the service manager has a chance to build a team. For example, the technicians in a shop may have a particular weakness or lack of interest in an area, such as front-end alignments. As a result, although a job may be for a technician to conduct general repairs, the manager may be looking for someone with strengths and interest in front-end alignments. This is why applicants should be asked as part of their application about the type of work they have performed and if they have any preferences. Obviously, in this example, an applicant with an interest and experience with front-end alignments would be given a higher ranking than a person who does not. Of course, in the interview the candidate would be asked to take a hands-on or performance test to demonstrate his/her ability to make a repair on an alignment machine. A person might say he/she is interested or good at anything to get the job. The performance test will tell if a tale is being told.

After ranking the candidates, the top people should be chosen and their references should be checked. This means a follow-up call should be made to the people who provided a written reference. The surprises that have been experienced by the authors and others from the follow-up calls are too numerous to mention. The most surprising was when an employer learned the reference did not know the candidate (who had already been hired). The candidate was using a false name and credentials. He was fired and the employer to this day does not know who the person was. In most cases, however, the references received over the phone are positive with the exception of the people who do not want to put something negative in writing for fear of a lawsuit. In this situation, the person may be asked if he/she owned a business or needed work done, whether or not he/she would hire or ask the candidate to do the work. A simple yes or no without any explanation is typically enough to get the information needed.

Conducting Interviews

The finalists need to be chosen from the top candidates and invited to the facility for an interview as well as a performance test. Before this occurs, however, there are several decisions that must be made and some ground rules laid. Most important, the person or people conducting the interview must keep in mind that the objective is to collect information from the candidates. They must be asked questions and their responses must be evaluated. If a candidate gives a wrong answer, it is not the responsibility of the interviewers to correct or argue with the candidate. The answer simply indicates the candidate's knowledge of the job is limited or incorrect. Another problem with interviews is that too often the people conducting the interview do most of the talking. This indicates that the interviewer is either not prepared or comfortable asking questions. Interviewers cannot instinctively pick out the best candidate based on how well the person listens or agrees with them. Interviewers must remember that, to build a winning team, they must take time to prepare relevant questions. Questions regarding marital status, sexual preference, and religious beliefs, among others, that are not directly related to the job are OFF LIMITS and actually illegal to ask. Questions must be limited to the job and the duties the candidate must perform. When in doubt about what to ask or not to ask, consult a human resource expert or search out publications such as the one shown in Figure 6-2.

When preparing for the interview, the first decision regards the person or people who will conduct or participate in the interviews. If a team is used, each member must be selected and trained in what questions are to be asked and what questions may not be asked. Managers should seek legal assistance in identifying the questions that may not be asked for legal reasons. In some cases, a question may be asked but it must be said in a specific way. Regardless of the person or people conducting the interview or giving a performance test, the same people must participate in all interviews and all applicants must be treated the same. They must be asked the same questions, given the same test, and given the same amount of time in the interview. If not, the interview cannot be valid, meaning it is not assessing what it is supposed to assess because one or

FIGURE 6-2 Publications such as this will help during the interview process.

more of the applicants received preferential treatment. The result may be that the best qualified candidate will not be hired.

The next decision is to determine who will be interviewed first, second, and last, that is, if the best candidate should be interviewed first or last. This is important because research has shown that the last person interviewed is more likely to be the one who is hired. Finally, the method to be used to conduct the interview needs to be decided. One method is a structured question-and-answer approach, while the other is the open-ended or conversation approach. Of course, a combination of both is preferred and the person or people conducting the interview must know the difference and be prepared to use them.

The structured interview is one in which the person is asked a question by the interviewer or someone on the interviewing team, as shown in Figure 6-3. The questions should come from the job tasks and duties in the job description. After a response is given to a question, the next question is asked. No follow-up questions should be asked unless some clarification is needed to understand the answer. The person or people conducting the interview should give a score on how well the candidate answered each question, as shown in Figure 6-4. The scores are usually on a scale of 1 to 10 or 1 to 7. When a candidate has completed the interview, the scoring sheets are collected and tallied by the person in charge of the interview. The people or person conducting the interview should not add up the scores for a candidate until all interviews have been conducted, because there is a temptation to rate people in relation to other scores.

FIGURE 6-3 An interview team conducts an interview.

FIGURE 6-4 Members of the interview team keep score on how well a candidate answered questions.

When all of the interviews are complete, the scores on each question should be added to get the total score. If more than one person was involved in the interview, their scores should be added and then divided by the number of people to get the average score. The highest average score gets the job.

The open-ended method of interview also should use questions that are the same for all candidates. Each question should receive a score by the interviewer or interviewers. The questions in this case allow the candidate to discuss or have a conversation about his/her knowledge about a subject, give opinions or ideas, and to talk about their experiences. For example, a candidate may be asked about his/her previous jobs, job tasks and duties performed, people with whom he/she worked, business relationships with former supervisors and coworkers, opinions about previous and current employers, reasons why they want to change jobs, personal likes and dislikes, and so on. In these cases, the responses are often surprising. The interviewer or interviewers must not react to a candidate's response regardless of whether he/she agrees or disagrees. As the authors can testify, some of the responses given by a candidate beg for a response. For example, a candidate may give a wrong response or make negative comments about all of their current and former supervisors, fellow workers, and employers. If this is the case, will the person have the same opinions if he/she is hired? If the responses are primarily negative, the person may be a negative force in the facility. The scores given to the candidate is a reason why more than one person should participate in the interview. One person may hear a response one way and another may have had a totally different reaction. As in the structured interview, the scores to the questions are added up and, if more than one person participates, the scores are averaged. The highest score wins.

Obviously, a combination of the structured and open-ended interview method is best. In the case of a technician, a performance test should be evaluated in the same manner. One option is that interviewers may place themselves in different locations to observe the work of the candidate, such as the setup of the equipment, the removal and replacement of parts, and so on. The objective of the test is to assess the person's skill level and work habits. As a result, only those duties that are relevant to the assessment need to be made. In other words, the performance observation does not need to review an entire job but only those needed to get an idea of the person's level of competence. Each candidate must be given the same test and, as the person performs the work, each observer should give a score. The scores should be added up and averaged to gain a final score.

After the interviews and performance test are conducted, the scores should be shared with the interview committee, if one was used. The scores may be looked at separately and also added together to give a grand-total score. If one of the interviews or the performance test is to be given a greater weight than the others before they are added together, the score simply should be multiplied by the weight to be given. For example, assume the total points to be earned were 70 points on each interview and the performance test and the grand total of points that could be earned would be 210. To give the performance test greater weight, the average points earned by a candidate could be multiplied by two. The person earning the most points would be hired, assuming that all qualifications were met and all references were acceptable.

Although the interview process should be an impersonal and efficient way to select a new employee, there may be problems. One, for example, is that some people may have a problem giving scores. In one case, a man gave all of the candidates a 10 because he liked them all. Obviously, this person should not be placed in such a position. In another case, one person on a committee had a favorite candidate. As a result, he gave all of the other candidates a low score (5 or below) and his favorite a 10. This gave his person the most points simply because of his bias. This person's scores should not have been counted. This is also one of the reasons why the number of points assigned to the questions should be limited to 5 or 7 points.

Making a Job Offer

All service facilities should have a clear-cut process to follow when a job offer is made. This is important because when an offer is made to a candidate, the facility is entering into a contract, even if it is not in writing. In some cases, employment contracts should be written and prepared, or at least reviewed, by an attorney who can make sure the employment rights of the worker and owner are stipulated in accordance with federal and state laws. In addition, the person having legal authority, such as the owner or person representing the owner, should make the offer and sign the employment contract. Naturally the contract depends on several conditions. Will the employee be full-time or part-time? Will the employee be expected to successfully complete a learning period before a position becomes permanent? If so, what are the conditions of the contract during the learning period? What is the obligation of the employer if the person's evaluation at the end of the learning period is negative? What differences should exist between staff exempt and nonexempt contracts? Finally, is a drug test required before the employment is finalized? One employer hired a new employee in a position with little supervision to service equipment and then had to fire him the next week because he failed the drug test.

Therefore, the job offer should not be made at the end of an interview. Although the details of a job offer, such as job title, pay, working hours, benefits, and so on, must be determined at the time the job was posted, these details must be confirmed before an offer is made. Next, the contract, which should be on file, should be reviewed so that it can be discussed with the candidate when the offer is made.

In addition, a snap offer at the end of an interview suggests that the service facility is desperate for workers and can scare the candidate away. At the same time, it is not fair to place a candidate on the spot. The candidate and facility often regret a decision simply because it was made in haste. People need time to think and talk to other people.

After all job interviews are conducted, an offer should be made promptly, such as within three or four days. Delays lose applicants. For this reason alone, the interviews should be conducted within a brief time period, such as a week, and the candidates should be told at the end of their interview when a decision will be made. When an offer is made, the person should be given a short period of time, such as 24 to 48 hours, to accept the job or sign a contract. For example, it is not unusual for a person to apply for a job, get an offer, and then go to his/her present employer to attempt to negotiate a pay increase or promotion. A time limit for a response to an offer is also important because, if the first person declines, the second person will have to be contacted. A lengthy delay implies to the second or third candidate that the job was offered to someone else who turned it down. This implication is not a good start to a new employer/employee relationship.

Follow-Up

The follow-up means that, after a contract is signed, everyone who was interviewed should be sent a letter of appreciation for the time taken to come into the facility. If the person's credentials and interview were impressive, he/she may be asked if the facility may keep their name on file in the case future jobs become available. The follow-up also should include a brief note sent to everyone who applied for the position. The letter should simply state that although their credentials were impressive, another person was selected to fill the position.

While these follow-up letters may seem unnecessary, they are important to the public relations and the reputation of the facility. When people apply for a position, they want to know that their application was received and reviewed. Although a letter stating the person was not accepted for an interview or not hired is not a positive experience, respectful treatment is usually always appreciated.

Final Thoughts

The recruitment and selection tasks should be designed to find and hire the best qualified person available to fill a position. One way to envision this process is to imagine that the job description and qualifications create a blueprint. Each position, therefore, should have a different blueprint. The recruitment task is to seek out people who fit the specific job blueprint. Then, the selection task has to choose the best qualified person by matching the credentials of each candidate to the job blueprint. If no one fits the blueprint, or if none of the candidates meets the specifications in the blueprint, then a new search has to be started.

In other words, a job blueprint could be compared to a mechanical blueprint for an electrical system. When an automobile is being repaired, the blueprint must be followed to see if all of the parts and wires exist and if the wires go to the right connections. If they do not, the electrical system may need major repairs or it may not be worth keeping. When a person is hired for a position, the missing parts or weaknesses found in the comparison of the person's credentials to the blueprint is where induction begins. If a person with weak credentials is hired, the induction and development tasks will have to assist the person strengthen his/her competencies. This will require training and time. While in the long run this may be worth it, the cost of the endeavor in time and money must be considered.

In addition, when managers are involved in the recruitment and selection tasks, they need to be reminded that before many people apply for a job, they have to think long and hard about it. After all, they have to go to some trouble preparing an application and résumé plus making copies of their credentials. In addition, people almost always feel threatened when they are being examined. In fact, because of the fear of being rejected and possible embarrassment, some current employees may not submit their names for consideration. This is why managers need to look actively for prospective candidates within the company.

Finally, how a facility and a manager handle the recruitment and selection tasks can have a considerable impact on their reputation with their employees and the people in the automobile service industry. If people believe they were treated unfairly when a job opened up, the bad feeling does not go away. If people believe they were not treated with respect or properly considered, positive advertising cannot overcome the negative impressions they will pass on to other people, including current and potential customers. In other words, for a number of reasons, when the recruiting and selecting of new employees is done poorly, it is bad for business.

Review Questions

1. What is the objective of the recruitment task?
2. What must be performed during the recruitment task?
3. What is the objective of the selection task?
4. Explain what the term best qualified means and why it is important to recruitment and selection.
5. What are the major duties to be performed in the selection task?
6. Explain the difference between the standard and open-ended interview.
7. Outline a recruitment and selection plan for a technician at an automobile service facility.

CHAPTER 7

INDUCTION AND PERSONNEL DEVELOPMENT

LEARNING OBJECTIVES

Upon reading this chapter, students should be able to:

- *Present the three most important responsibilities of a manager and explain why they are important.*
- *State the purpose of induction.*
- *Describe what must be contained in job and nonjob induction programs and why.*
- *Explain the difference between a formal and an informal leader.*
- *Explain the relevance of mother duck imprinting to management.*

- *State the purpose of development.*
- *Present the two objectives for personnel development programs.*
- *Present the differences in the types of employee personnel development programs.*
- *Identify and describe the four subplans to be included in a development program plan.*
- *Explain how ASE certifications can play a role in personnel development.*

Introduction

In addition to the recruitment and selection tasks, the service manager has three other personnel job tasks to perform. These tasks are induction, development, and appraisal. Like recruitment and selection, the duties to be performed in the induction, development, and appraisal tasks have been standard practice for over 50 years. The purpose of this chapter is to discuss the duties to be performed in the induction and development of employees. In this discussion, when the duties are explained, possible alternatives are pointed out when appropriate.

Induction is required when new personnel are hired or when someone in the company is promoted to a new position. The induction of a new employee lasts until an employee has completed a learning period in a provisional, or temporary, position. Upon proving him or herself competent, the person moves from the provisional position to a permanent position. On the other hand, the duties required in personnel development never end. They are always important for all employees, including the service manager. They begin when an employee is hired and end when a person's employment has concluded.

Development programs are conducted for both the advancement of an employee's knowledge and skill and also for company improvements. These programs are especially important in the automobile service industry, which faces constant change. For instance, every year automobiles contain new technological advancements, which create additional demands on technicians and facilities that provide automotive services to the public. The changes on the horizon for the industry appear to be vast with fuel cells, hybrid electric vehicle battery technology (see Figure 7-1), and onboard information technology that require use of special tools/computers to diagnose such as that seen in Figure 7-2. Each of these changes promises to proceed at an alarming pace and automotive service facilities cannot (and do not) stay in business if they and their employees are not able to stay current.

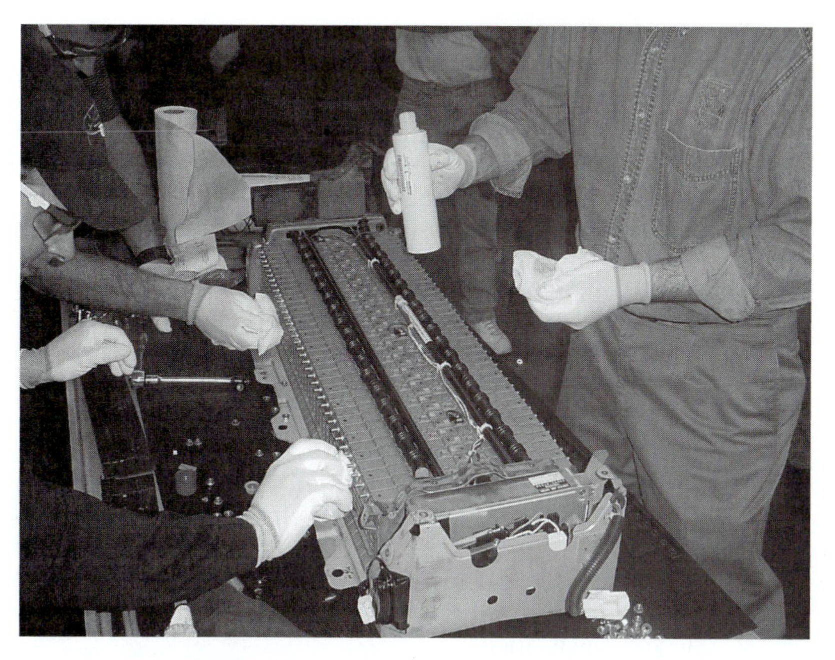

FIGURE 7-1 Hybrid electric vehicle battery.

FIGURE 7-2 Diagnosis of onboard information technology requires the use of special tools and computers.

The Mother Duck

When ducklings hatch, the first living being (human or animal) they see becomes their mother. They follow this *being* around wherever it goes. The *being* becomes the little duck's mother, even though it may not be a duck. Often times, a student on a college campus learns about this phenomenon and *hatches a duck egg*. The duckling then follows the student around campus due to what psychologists call *imprinting*. Likewise, when new employees report to a job, the person who inducts them usually becomes their mother duck. When new employees have a problem or need advice, they go to the mother duck. When a person becomes the mother duck to all new employees, this person becomes the authority. The person with authority becomes the leader and, all too often, the manager is not seen as the leader. Therefore, the manager should be the first person the new employee meets. Furthermore, the manager must have an induction program prepared for the new employee and, by example, show he/she is *mother duck* in charge. When the new employee has any questions or concerns, the person should go to the manager for assistance.

An example of a mother duck problem occurred at an automobile dealership. In this case, the managers were quite frustrated with the practices and attitudes of the parts specialists. They tried to come up with different approaches to change the parts specialists' attitudes and work habits, but were not successful. When the managers asked about the induction of the new employees in the parts department when they were hired, they found that the person who took care of that responsibility was a secretary in the parts department. As in many similar cases, this person was a long-term employee and had a reputation of running the place. When asked if there was any similarity between her attitude and work habits and the practices and attitudes of the parts specialists, the light bulb went on and the managers could see the relationship. Because she had worked at the facility for many, many years and then finally announced her retirement, the managers decided to wait for her departure before setting up and managing a formal induction program, which followed the recommendations of this text. Of course, they also had to offer development programs to shift the attitudes and habits of current workers (as discussed later) who were taught by the retired mother duck. Eventually, the secretary's influence was reversed, the attitudes and habits of the employees in the parts department changed, and it became the professional operation the owners wanted.

Another problem that can take place is when a new employee is allowed to wander around the facility to get to know everybody on his/her own. In these cases, the manager allows the *informal leader* to become the mother duck. While the manager is the formal leader, the person hired to lead the unit, an informal leader is the person the employee will go to for advice. This is often a problem since the informal leader may not be a positive force in the organization or a supporter of the manager. Ideally, the manager should be the formal and informal leader; however, the manager actually may have to compete for the role of informal leader. In other words, some employees want to be the leader and, although they do not qualify for the manager's position, they may have the personality to fill the role of the informal leader. Therefore, when a new employee is left on his/her own to meet everyone, an informal leader can take over and induct the new employee to "the way we (I) run things around here." In these cases, the manager is typically in trouble since the informal leader can set informal company policy and procedures that control worker attitudes and habits. Observations over the years have shown that these informal policies can create a self-regulating culture. New managers may be hired to change operations, but the culture set by the workers keeps everything the same. In some cases, the prolonged negative influences of a self-regulating culture can shut down a business. The bottom line is that the informal leader cannot become the mother duck.

If a manager cannot fill the role of the informal leader and mother duck because of time constraints, personality limitations, or simply too many employees to handle, he/she should identify and prepare a mature employee who supports the facility to serve in this position. For example, when a new employee reports to work, the employee should report to the manager. The manager should explain the purpose of the induction program and then have the new employee's immediate supervisor take the responsibility for becoming the *mother duck*. In this setup, the manager is the supervisor's mother duck and creates a boss's boss imprint for the new employee. More specifically, if the facility uses the team approach, the manager may have the team leader serve as the mother duck and informal leader of the technicians on the person's team. When the new employee reports to work on the first day, the person would first meet with the manager and then the team leader. The team leader would then review the person's induction program in detail, as shown in Figure 7-3. This should close the door on just any employee who might aspire to become the informal leader and helps the new employee, as well as the other employees, know that the manager is in charge.

Induction Programs

Induction programs are not only prepared for new employees but also for existing employees who have been awarded a new position in the company. As explained in the last chapter, a new employee is not likely to perfectly fit a job description or blueprint. As a result, the person is placed in a temporary position for a learning period. During the learning period,

FIGURE 7-3 It is the responsibility of the team leader (left) to review the new employee's (right) induction program.

an induction program is provided to assist a new employee to fill in the blueprint's job expectations. When these expectations are met, the employee is advanced from the temporary learning position to a permanent position in the company. These learning periods vary from one person to another because people have different educational backgrounds and various types, levels, and years of work experience. For this reason, induction programs are customized to fit the new employee. Consequently, the ultimate purpose of an induction program is to provide a new employee with on-the-job assistance over a learning period in order to qualify them for a full-time permanent position.

More specifically, in the learning period, the employee must exhibit the ability to perform all tasks and duties presented in the job description. In addition, the program must include at least the following non-job-related support:

- Instruction on operational procedures (Chapter 4)
- Review of the operation manual (Chapter 4)
- Discussions of company practices and policies (Chapter 4)
- Introductions to personnel and their position in the organization
- Formal tours of the facility and introduction to their work area
- A checklist of the use of equipment for safety purposes

Finally, another reason for having an induction period for a new employee regards the need to adjust to a new environment. In other words, when entering a new environment, the person must interact with new people and get used to different routines, noises, smells, and so on.

These are all strange and it takes time to get used to them. Some people adjust more quickly than others and the support given by the employer can influence a person's on-the-job performance.

Using Performance Standards

Performance standards are targets that an employee is to meet in an induction program or any evaluation of his/her knowledge and ability. When a manager reviews the employee during and at the end of the induction period, the standards inform the new employee what he/she will be expected to do and how well he/she will be expected to do it. Performance standards also are used when a manager performs an employee's annual review, as discussed in the next chapter.

A responsibility of a manager is to make sure each employee understands the standards that will be used to measure his/her performance. For managers, these standards should serve to maintain and improve effective and efficient operations, which should be linked to the standards the owners and/or upper managers use to evaluate them. For instance, if a manager is held to specific minimum outputs by the owners, he/she will not likely meet them if the people who work for them do have specific standards to meet that output. When standards for employees or managers do not exist, then each worker is basically permitted to set his/her own standards. This is extremely dangerous and career threatening to a manager.

Performance standards should be given to a new employee during an orientation period. In an orientation period, the employee usually does not do any work so that he/she can fill out tax forms, set up their work space, be introduced to other employees, receive a tour of the facility, and so on. In addition, during the orientation period, the manager should explain how the induction program works and how it is intended to help the employee meet the required standards of performance. Next, the manager or supervisor should inform the employee of the specific performance targets to be met. The use of the targets should assure the new employee that his/her evaluation will be based on his/her performances, that new targets will not be added, and that evaluations will be based on personalities. The person also must know that when the performance standards are met, he/she will be promoted to a permanent employee status, which will earn him/her benefits (vacation days, sick time allowances, medical, retirement, etc.) and perhaps an increase in pay.

Setting and Checking Performance Targets

As noted, the performance standards should present specific targets an employee is to meet. These targets for employees at a service facility will vary by job title, expertise and even specialty. For example, a flat-rate B technician may have a weekly target of 50 flat-rate hours that is based on his/her expertise. A service consultant may have a target of trying to earn two flat-rate hours per car on average each week. These two targets are different but actually complement each other, as they both help the service facility meet its objective to earn a profit. In other words, the targets should be an integral part of the business and connected to company objectives. To help a manager select the appropriate targets for an employee, the target must indicate:

- When the task or duty must be performed. (As discussed in the next chapter, this is referred to as the condition.)
- What tasks or duties the employee must perform. (As discussed the next chapter, this is called the performance.)
- How well the employee must perform the task or duty to be considered acceptable as defined by the effectiveness and efficiency. (As discussed in the next chapter, this is called the standard.)

Note that the term *perform* is used. This term is also used throughout the next chapter on evaluation. The question should always be whether the person can perform the job and get it done. Performance standards and targets can be in relation to technical work (done to a level that will create profit as well as quality work), interactions with customers (to a level that will ensure they are satisfied) and fellow workers (to promote a team atmosphere), professional appearance, attitudes, habits, safety, basic employability skills (such as getting to work on time), following instructions, and so on.

Benchmarks

Final performance standards are the targets a new employee must meet by the end of his/her induction period. During the induction period, targets also can be set for the end of a time period (such as every Friday) or for specific work the employee performs during the induction period (such as a review after the technician's first engine job, as shown in Figure 7-4). The periodic targets should be used to help the manager determine whether the employee is making adequate progress toward meeting the final performance target. These periodic targets may be referred to as benchmarks, performance benchmarks, or benchmark targets. When the employee has met the benchmark target, the employee should continue on the induction program as planned. When the employee does not meet the benchmark, the manager should adjust the plan to provide assistance so that the employee can meet the target. If an employee is given additional assistance and still cannot meet the benchmark target (perhaps the employee overstated his/her ability during the job interview process), then the employee should be counseled into another new job if available or terminated.

The time periods set for the periodic and final benchmark reviews depend on the employee and job. A benchmark review could be every two weeks during a 12-week induction program for a less complicated labor job, or every four weeks in a 16-week induction period for a technician with considerable experience in more complex position. Another option

FIGURE 7-4 A target, such as the amount of time to finish the last step of an engine job, can determine whether adequate progress is being made to meet a desired benchmark.

is to schedule benchmark targets on the performance of an activity or completion of a job, such as an oil change, brake job, tune-up, installation of an engine, and so on. In some cases, the manager may require that a new employee be assigned certain jobs within a specific time period, such as a series of oil changes within the first week of employment. The reviews are then conducted after each job is completed or series of smaller jobs are completed. The intention is to become aware of problems and correct them before the new employee develops bad habits and repeats the same error on the next job. As the induction period progresses, new and more substantive targets should be set. For example, oil changes may be the first benchmark followed by tire installations, tune-ups, and so on. Naturally, the employee's abilities and experiences will determine where to start and to what extent the supervisor will observe the task duties and work details. In other words, an A-level technician with 15 years of work experience would not be reviewed for oil change proficiency, but the benchmarks would focus on more difficult tasks.

To illustrate the importance of benchmarks, a new employee at the authors' service facility claimed to be an expert on engine repair and, from all appearances, seemed to know what he was doing. After the reassembly of his first engine job, the authors conducted a benchmark review. In preparing for the review, they realized almost all of the entire new bolt inventory kept at the service facility was gone. They then learned that the new employee used them during reassembly of the engine. When asked why he replaced every bolt he removed, he stated he had put the bolts from the engine in a box. Because it would take too much time to sort them out of the box, he decided to replace them with new ones. For the authors, this meant that the cost to replace every bolt in the engine decreased the profit earned on the job. The review also revealed that some of the bolts he used were not the correct strength. The engine had to be disassembled so bolts with the correct strength could be installed. As a result, the entire job had to be reworked and the cost was absorbed by the owners (authors) of the service facility. Of course, the problem should have been caught sooner by observing this person (setting benchmarks) during the engine repair process and not after the job was completed. Had there not been a performance review after his first engine job was finished, he would have repeated the practice on the next engine job. Obviously, this new employee did not truly possess the expertise, experience, or qualifications he claimed to have in the interview. Additional on-the-job training and supervision were provided.

When Induction Begins

Induction programs actually begin in recruitment when the job announcement is released and continues through the interview process. In other words, the job announcement should tell every applicant, including the person who is to become the new employee, something about the job and the company. For example, in the sample advertisement in the last chapter for a technician, it stated the facility was looking for a B technician. This told the person, who would be the eventual employee, the level of expertise expected by the facility. In addition, the advertisement indicated that benefits were available. This suggests to applicants that the owner is concerned about the health of their employees. The ad also mentioned that uniforms were provided. An owner who expected employees to present a professional appearance to the public would supply uniforms. If a person wished to work at a messy shop with dirty clothes and sloppy work area, this would not be a likely possibility.

The interviews also offer opportunities to inform the person who would be chosen to fill the position about the expectations of the facility. Specifically, the owner's expectations, job qualifications, and performance standards should be discussed with each candidate, one of whom will be hired. Numerous other job features, such as being on time, can be talked about. During the interview, notes should be taken on each person's reactions and comments that would be helpful in the induction program. These notes will remind the manager and supervisor of some of the performances to be discussed and checked in the induction program. For example, if a person had a weak or hesitant answer about a job task, then the person may need assistance on the conduct of the duties in that task. If a person's personal appearance in the interview was less than acceptable at the facility or if he or she explained that a current employer did not care about appearance, then the standards set for employee appearances will have to be discussed and checked during the induction period.

Job Description Tasks, Duties, and Targets

The most common source of information used in an induction program is likely the job description. The tasks and duties in a description of a job should be used to create job performance standards and induction targets for new employees. The manager must decide if there are different skill levels for conducting a task and duty and, if so, at what level should the new employee perform them. The manager must also recognize the differences among technician grades; for example, if all techs had a common task, the proficiency level for a C-tech target would be different from that of an A-tech target.

Based on a comparison of the performance targets set for the new employee and the employee's background (education and work experience) and information recorded at the time of the employee's interview (which would include the performance tests), an induction program would be prepared. The comparison would then be reviewed in detail with the employee at the orientation meeting. From the personal experiences of the authors, this review is likely the most critical part of the induction program and should be used in setting the benchmarks. For example, a job description for a service consultant, which was used by the authors at their service facility, is shown in Figure 7-5. On the first day that a new service consultant reported to work, the service manager went over the each of the tasks listed on the job description. In the review, the manager gave him an honest appraisal of his potential to perform the tasks based on the interview. The service manager then went over the assistance that would be provided to bring the person's performance up to the expected level and the use of scheduled benchmark reviews.

As an illustration in the use of a job task and performance review, assume Rob is a new service consultant who had a position as an office helper and that he assisted the management staff. In this position, he had never been required to greet customers (second task in Figure 7-5). The service manager, therefore, must provide instruction with an example of how Rob would have to greet customers. Rob would then be advised that, during the learning period, he would be observed to see how well he performed the greetings. These observations would be discussed with him at his first benchmark review.

In other words, in the review of duties in a job description, when possible they must be reviewed in detail. For example, Task 11 requires the service consultant to close the building. The duties for this task should be detailed as shown in Figure 7-6 and on a checklist. The service manager must go over each of the duties, walk through them with the new employee, and then observe the new service consultant close the facility at the end of the day. During the learning period, the service manager must check to see if the new service consultant performed all of the duties as expected, such as checking the thermostats and lights before leaving the building for the day. While this may seem unnecessary, on one occasion the authors found that some of the closing duties

Job Description for a Service Consultant
Position Description

Service Consultant

1) Opens shop at 7:30 a.m.
2 Greets customers, answers phone calls, provides information, makes appointments, calls customers for approval of work on vehicles, calls customers when jobs are completed, and places follow-up calls after repairs are made.
3) Prepares customer invoices, reviews parts and labor charges with customer, and receives payment from customers (cash, check or credit card).
4) Makes arrangements for customer shuttle or comfort if waiting at the service facility.
5) Prepares estimates, Repair Orders (RO), and computerized invoices (IN), and maintains the appointment book and customer status sheet.
6) Communicates with the technicians to ensure service work is completed in a timely fashion.
7) Communicates with technicians as needed to prepare estimates.
8) Assists as needed to help order parts, receive parts orders, and check parts invoices to ensure the charges are accurate.
9) Maintains inventory of office supplies including supplies for computers, photocopier, fax, credit card machine, printers, forms, reports, and other materials as needed.
10) Makes daily deposit to the bank and prepares daily report for management.
11) Closes building at the end of the day and checks customer cars in parking lot.
12) Other duties as assigned by management.

FIGURE 7-5 Service consultant job description.

were not performed. When the service consultant realized that his performance was being checked, he started to pay greater attention to the details assigned to the job.

When the service manager checks on the performance of an employee (new or existing), notes must be kept on exceptional, good, poor, and neglected/bad performances. The exceptional and neglected/bad performances should immediately be called to the attention of the employee. At the benchmark review, the service manager should cover all of his/her notes with the new employee. This review should include the poor performances first and then close with the best performances for the sake of maintaining good morale. In other words, closing a meeting on

Task #11 Duties—Closing Building
CLOSING RENRAG AUTO REPAIR

IF WINTER OR FRIDAY NIGHT—put all company automobiles, including the customer shuttle, in the back bay.

BACK BAY
1) IF WINTER, turn thermostat down to 55 degrees (if it is to go below freezing) or OFF if it is to stay above freezing.
2) Close bay doors (make sure motor turns off by looking at spindle in ceiling).
3) Make sure all drop lights are off.
4) Check exit door (back wall) and make sure BAR is across the door.
5) Check water faucet (with hose) and make sure it is OFF.
6) Turn off bathroom light.
7) Turn off bay lights by compressor room.
8) Enter compressor room and turn off compressor.
9) Turn off compressor room light.

SIDE LUBE BAY
1) Go to lube bay and close and lock two bay doors.
2) Turn off tire-balancing machine.
3) Make sure all droplights are off.
4) IF WINTER – turn thermostat down to 55 degrees if waste oil heater is on.
5) Turn off lights to lube bay and upstairs light.
6) Go to internal bay lift and turn lights off.
7) Go to bay with alignment machine.

ALIGNMENT BAY
1) Close bay doors.
2) Make sure side exit door is closed.
3) Make sure all drop lights are off.
4) Turn off alignment machine.
5) IF WINTER, turn thermostat down to 55 degrees (if it is to go below freezing) or OFF if it is to stay above freezing.
6) Turn off light in bathroom – leave door open.
7) Turn off bay lights.
8) Go into front office and enter interior office.

INTERIOR OFFICE
1) Turn off all computers and printers.
2) Cover computers with protective covers.
3) Make sure inspection sticker box is locked.
4) Turn off lights.

(continued)

FIGURE 7-6 The duties for Task #11 as found on the service consultant's job description (Figure 7–5).

OUTSIDE OFFICE & WAITING ROOM

1) Turn off all computers, printer, and TV security monitor.
2) Switch answering machine to position "B."
3) Turn off photocopy machine (leave fax ON).
4) Turn off radio and TV.
5) Turn off lighted OPEN sign.
6) Turn "CLOSED/OPEN" sign around to show "closed."
7) Make sure coffee pot is OFF.
8) IF WINTER, turn thermostat down to 55 degrees (if it is to go below freezing) or OFF if it is to stay above freezing.
9) IF SUMMER, turn AC off.
10) Arrange furniture and magazines; put cups, etc., in trash.
11) Lock front door.
12) Set alarm, turn off lights, leave, and lock door.

PARKING LOT – CUSTOMER VEHICLE INVENTORY

1) Check all vehicles in lot and make sure they are locked.
2) Record all customer automobiles left overnight and placed in the building or left in the parking lot on the customer automobile inventory sheet.
3) Note any damages or missing parts on inventory.
4) Enter date and time the inventory was taken.

FIGURE 7-6 *(continued)*

what the employee did wrong or poorly is not recommended. After a performance review is over, a memo may be prepared to compliment the person, to clearly define a task and/or duty that seemed to be confusing to the employee, and to issue a warning on any duty on which the person's performance was not adequate and that must be improved by the end of the induction period.

Induction and Operational Procedures

As explained in Chapter 4, the operational procedures describe how the different work units are to work together to meet the objectives of the company. This includes a detailed description of what each work unit is to do and how work is to flow between them, such as the flow of repair orders. In addition, the operational procedures should provide the company's expectations with respect to the induction of new employees.

Consequently, the induction program for new employees must include a formal review of the company's operational procedures that pertain to the new employee's unit and job. All too often, procedures are ignored, casually mentioned, or new employees must discover them for themselves. Although a formal review of procedures may seem unnecessary because it takes away from the time new employees can get to work, it will pay off in the long run. The reason is that new employees have to ask around about the way to do things, and opportunities are opened for wannabe informal leaders and even the hazing of a new person. The total time lost is greater than the time it would take for the manager to personally review or delegate someone to review the procedures with the new employee.

Induction and the Operation Manual

As also explained in Chapter 4, the operational procedures are used to prepare an operation manual for each work unit. In these procedures, the company's objectives and performance expectations set forth for the service department or unit are stated. As a result, they must be made known to all employees when they are hired. The operation manual specifies what the service department is to do and how well the employees in the department are to do it.

Because the manual is intended to obtain a consistency in the performances of a work unit, such as the service department, all new employees must be given a copy. The manager should give the new employee time to read it and then either he/she or an assigned delegate, such as a team leader, should personally review the information in the manual with him/her. This will ensure the new employee understands all of the directives and activities to be performed each day, week, and month. For example, the manual should include instructions on safety procedures, such as personal safety, safety of others in the shop, safe equipment use, and so on. Covering these directives in detail is not just a good idea of the authors but an expectation of the company's insurance carrier. Other information that would be covered includes specifics on the processing of work, use of inventory, servicing equipment, and so on.

In particular, the manager must point out the performance standards for the service department and explain how they are used to assess employee performance. For example, in the case of a technician, the manual would indicate the different places that work orders are to be placed during the service process. The best way to cover this type of content would be for the service manager to walk through the process with the new employee. The manager must then explain that when the new employee fails to put work orders in the correct place, the efficiency and effectiveness of the facility would be harmed. In turn, the manager can explain that a warning would be issued that could lead to the extension of the learning period (probation) or even termination.

Induction and Company Practices and Policies

As discussed in Chapter 4, company policies should be in a manual to indicate what employees can expect from the employers and vice versa. In particular, the policies related to personnel matters must be carefully reviewed. While some policies are best reviewed with new employees by the business office—for example, pay dates, taxes and tax deductions, benefits, charges for sick days, vacation, and so on—others should be handled by the manager. For example, overtime is both a policy (because it has to do with wages and state law) and an operational procedure (because it has to be authorized at the discretion of the service manager). For instance, employees must understand that just because a person works beyond quitting time does not mean overtime is automatically paid but it must have the manager's authorization. In other words, the owners set the policies and rules and regulations are created for the managers to carry out as intended by the policy. Another example of a company policy in an operation manual states when the company opens for business. The manager may add some rules and regulations to this policy by directing where employees should be when the business opens. In other words, the manager will not want employees pulling into the parking lot at the start of the business day; they should be at their workstation and ready to begin work.

In addition, as discussed in Chapter 4, some company policies also are based on public laws, some of which are quite complex. The manager cannot assume that new employees are familiar with any of these laws, even if they have worked at another facility for a number of years. Some of the legal policies are quite complicated; and in some cases, employees may have to sign a statement indicating they read and understand the company policy, such as a policy on sexual harassment, drug abuse, and so on. In addition, new employees should be informed about their unemployment benefits and **workman's compensation** as all too often they have obtained incorrect information, especially in regard to unemployment benefits at the end of the learning period (probation). The manager also must check to make sure employees are informed of other legal policies. For example, in Pennsylvania, laws on automobile safety inspection must be covered even if the person is licensed as a state safety inspector. The manager cannot take anything for granted and must always make sure new employees as well as all members in his/her department are correctly informed.

Induction and Introductions to Personnel

While the introduction to the owners and employees does not take much time, it is important and should take a formal approach. First, the manager should set up an appointed time to properly present the new employee to the owners or, in a large organization, to the manager's supervisor. These introductions should not take an *off-the-cuff* attitude and catch the people unprepared to offer appropriate greetings. These meetings must grant the owners and supervisors the respect they deserve. In addition, the new employee must appreciate the position of the person whom he/she is meeting.

When meeting future coworkers, the manager or the new employee's supervisor should make sure that the new employee is properly introduced to everyone. If the new employee is to become a member of the manager's team, the person must be given a proper introduction and then allowed time for a brief conversation. Informal introductions open opportunities for informal leader *wannabes* with improper ideas, such as hazing.

Induction and Formal Tours

A tour of the facility is absolutely necessary and should be conducted in the orientation. Even though the new employee may have had a brief tour at the time of the interview, a more in-depth tour is needed. On the tour, important features need to be pointed out, such as the location of safety equipment like fire extinguishers, eye wash equipment, a first-aid kit, emergency exits, electric kill switches, and so on. Of course, the new employee's workstation also must be examined and the company-owned equipment (and supplies) identified and inspected for the benefit of the new employee as well as the owner. The new employee must recognize what the company owns and know that it operates properly, and must not assume it was left by a former employee.

Induction and Equipment Check-Off

Managers cannot take for granted that a new employee knows how to operate a machine, such as the lift shown in Figure 7-7, even if the person says he/she does. For example, the authors owned some of the latest

FIGURE 7-7 Equipment check-off is an important step in the induction process in order to determine if an employee knows how to operate a machine, such as the lift shown here.

equipment models and frequently found that even though a new employee may have used such equipment, they were not familiar with it. In this part of the program, the manager (or person conducting the equipment check-off, such as a shop supervisor) should require the new employee to operate the machine. After operating the new equipment, the employee would either be checked off as being able to operate it properly or offered instruction on how to use the equipment safely and properly. When demonstrating that he/she knows how to operate the equipment, the check-off sheet would be placed in the employee's file. Preferably, employees who are the most familiar with a piece of equipment should be asked to check off a new employee. Although it takes away from production to check off a new employee on equipment, it is better than the alternative, such as broken equipment, damaged vehicles, or injured people.

Time Limit for Induction

The service manager must make the decision about the length of the learning period on a case-by-case basis. A short learning period may put the new employee at a disadvantage, as he/she will not be able exhibit all of the desired performances. At the same time, a long induction period postpones the time the person may obtain the advantages from being classified as a permanent employee. More important than the length (in some states, 90 days is considered the probationary period, during which an employee can be terminated and not be eligible to receive unemployment benefits) is the opportunity for a manager to have adequate time to fully review the tasks and duty performances the employee is expected to achieve. However, a learning period could end earlier than expected or a decision could be made to extend the learning period or terminate the employee. In all cases, the service manager must record the events to document the process from start to finish: "Document, Document, Document."

The End of Induction

When a person has completed an induction program and the performance targets are met, he/she becomes a permanent regular employee. If the performance targets are not met, the person may be:

- Put on a warning status with pending termination if performances are not improved.
- Given a new program plan to fit the new employee's abilities or lack thereof (pay may need to be adjusted if the employee's skills are not to the level implied in the interview).
- Terminated.

The successful completion of the induction program essentially means the person has proven his/her competence and will no longer be subjected to questions about his/her ability. The person's pay should be adjusted to reflect his/her true ability and job benefits should be granted, such as sick days, vacation days, receiving a uniform and laundry services, health and other insurance offered by the employer, eligibility for unemployment compensation (typically, the time period is dictated by state law), retirement investment opportunities, and so on. In other words, the move from a temporary to permanent position should not simply be a ceremonial recognition. When it offers benefits and a pay increase, the induction program becomes very serious for the employee and the employer.

Development

Personnel development programs apply to all employees, including people who have just been hired and are in an induction program. The purpose of the personnel development task is to ensure the company and its employees are up to date in the automobile service business. Being up-to-date is important because of the changes in the automobiles, such as new computer systems, hybrid powertrains (such as the hybrid transmission seen in Figure 7-8), alternative fuels (such as the CNG fueling station seen in Figure 7-9), and so on. For management and owners, staying current is a major concern because it is necessary in order for the business to be profitable.

FIGURE 7-8 Example of a hybrid transmission.

FIGURE 7-9 A CNG fueling station.

Probably the single most important factor for a successful development program is the commitment of the owners and managers. This means many things but, at a minimum, it includes funding the program, allowing employees time off to attend programs, and providing support (tools, equipment, information) when the person returns from the program. Employees must believe that management thinks staying current is important. This, of course, also means that owners and managers must stay current. Therefore, owners and managers must be willing to invest some of their own time, effort, and even a set amount of business earnings in development. Finally, the importance assigned to personnel development must be reinforced through some form of recognition or reward to employees who are making the effort to stay current. These acknowledgments are usually not expensive but have proven to be effective motivators.

A plan for a development program includes at least five subplans or parts, which are:

1. Program objective
2. Content
3. Delivery
4. Evaluation of employee performance
5. Evaluation of program effectiveness

A common mistake made in development is to participate in or provide development services to an employee when it becomes available instead of planning for what the business needs. In other words, because the planning process is usually not a high priority at a service facility, when an opportunity comes along, the manager may make a decision to take it. In this haste, the development program may or may not be what the business needs. In fact, many of these opportunities are usually expensive and often do not completely fill the needs of the facility. Therefore, development program plans should be prepared before entering into any agreements to invest money and time on any program.

Two Objectives for Development Programs

All too often, development is thought of in terms of a course or workshop taken by an employee. These often focus on a specific task, a piece of equipment, a process, or a law related to the employee's job. This is unfortunate, because the development task at progressive corporations has taken a much broader scope. Specifically, to be current, a development program should have one plan for the business that relates to the strategic business plan and a second plan for employees. Both involve specific objectives, content, delivery, and evaluations. However, although each has a different focus, they support each other because their purpose is to benefit the facility.

The objective of a program designed for the business is to keep employees informed and up-to-date on changes that affect the operation of the facility. For example, employees must be informed about revisions in state laws and new equipment, and these changes must be integrated into the daily operations of the facility. For example, because of the federally required emission program launched in Pennsylvania, progressive service facilities prepared a development plan for the addition of the new machines, such as the machine seen in the background of Figure 7-10.

FIGURE 7-10 Emission inspection machine.

This meant the integration of the equipment and federal mandates into their facility's operations and the training of their technicians before the inspections were required. Facilities that waited until after the emission inspections were required to be implemented ended up in turmoil. They risked losing their customer base to competitors who were prepared for the change before the program's launch.

With respect to the development programs for employees, they should relate directly to the employee's job tasks and duties. In order for a facility to keep up-to-date with the changes taking place in the environment, the employees must be prepared to properly handle them. For instance, in the Pennsylvania Emission Program launch, while the facilities were busy preparing for changes in their operations, the technicians and some managers had to be trained to learn about the emission laws and the use of the computerized equipment. Then they had to obtain a license in accordance with state law so they could perform emission inspections. Once they were licensed, the progressive companies were ready to immediately conduct customer inspections when required. As a result, the employee development program in the progressive companies was clearly related to the business development program. In addition, the business and employee development programs influenced the strategic business plan of the facility because capital had to be invested to buy the equipment and money had to be spent to train employees. In response, the progressive service facilities that planned for the change benefited from an immediate generation of profit. Keeping their businesses competitive had two rather than one objective for development.

Program Content

The content and purpose of a development program for a business must be connected to the strategic business plan. As discussed in the first part of this book, the strategic plan must be reviewed annually and modified as changes occur in the environment, such as the emission inspection requirements discussed earlier. Changes in laws, products, services, organizational structures, and so on lead to the modification of goals and objectives. These changes, in turn, lead to changes in procedures and the operation manuals. Before the changes can be put into place, however, a development program must inform the employees who will be affected and how they will have to make adjustments.

In some cases, development programs for a business's benefit require the manager to obtain information to understand what is needed. For example, information collected from a customer survey indicated they experienced confusion about a new service offered by the facility. This kind of information suggests the need for a development program for the service consultants and others involved, possibly on ways to improve their communication with customers, more information about the new service being offered, or a restatement of the way the special service is presented. This approach is much more effective than issuing criticisms and threatening reprimands, which provide no help to the employees.

With respect to meeting the objectives of development programs for employees, the content typically will focus on the improvement or advancement of an employee's job knowledge and skill. The content is usually about the processes, specific tasks, and even equipment to be used by the employee to perform the work in question. In one example, when a new automotive license is required, such as an emission inspection license, a development program would enter the technicians into a program to prepare them to pass the exams. In another example, a development effort might be helpful when the time taken to conduct an oil change in a quick-lube operation doubled. These two examples would require different employee programs with different purposes. The first is to educate an employee in order for him/her to take a test and the other is to improve employee performances.

In the first example, the manager would seek out a program that would add to an employee's personal knowledge. In this case, an employee would attend a program that would inform him/her about the definitions, measurements, repairs, diagnostic routines, and so on related to automobile emissions. The person also should be shown how to perform the inspection and then given a sample written test to measure his/her knowledge. At the end of the program, the technician should be prepared to pass the state exam.

In the second example, the manager has to increase the daily number of oil changes. The low number of changes appears to be a result of the inability of the employees to get the job done. In response, the manager must provide an *in-service* development program. In-service programs are designed to overcome a deficiency or correct a problem by offering training at the service facility to improve employee skill levels. Because an attitude problem also could be an underlying issue, the manager or trainer would cover company work expectations and values to put the employees in the right frame of mind. In other words, the skill improvement would be brought about through physical training, such as that used by the military, that is, by the supervised repetition of the work to bring the employees' performances up to standards. Such an in-service skill development program would teach and review the skill, permit the employee to practice it, and then determine if the person can perform it at an acceptable level.

Program Delivery

There are many methods for delivering a development program. Programs that focus on changes within a specific company, such as the business plan, operations, policies, and so on, are typically provided by a company employee who can directly identify with the need for the change and answer questions. Programs whose objective is to advance or improve an employee's job knowledge are often provided by an outside source, such as an educational institution, manufacturers, professional workshop series, consultants, and so on. Likewise, skill development programs may be provided by an outside source, such as the program seen in

FIGURE 7-11 Skill development may be provided by an outside source.

Figure 7-11, sometimes even on site. However, programs also can be delivered on a more informal basis by a supervisor or manager, as at some service facilities.

With respect to in-service programs, the type of deficiency may dictate the source for the program delivery. For instance, some consultants offer a sales training and education program with follow-ups to monitor changes in performance. Although such a source may be expensive, they are often quite effective. At the same time, a facility may not wish to reveal its internal problems. In these cases, a consultant may be hired to assist in the development of the content and the delivery method but not actually make the delivery.

Regardless of the source of the delivery of a development program, a manager should seek out alternatives to meet an objective. In many cases, if possible, an employee should not deliver a development program because it takes the person away from their daily job tasks. In addition, the extra responsibilities could affect their job performance. When an outside source is hired to deliver a development program, a contract must carefully spell out what is to be offered and, especially, the expected outcomes. The contract must indicate:

- To whom, when, where, and how the program will be offered?
- What is to be accomplished?
- What materials will be provided by the trainer?
- What performance reports will be provided?

- Who will do the training?
- What the cost will be?

When this information is received by the manager, it can be compared to the objectives and content/training proposed in the plan.

When a facility sends an employee to a program put on by an outside source, the manager must have a clear quote on the cost and determine who will pay the charges. For example, if the employer pays the fee, should the employee be committed to staying with the employer for a specified period of time? In one case, the authors' service facility paid for an assistant manager to attend classes to obtain the state safety inspection license. Shortly after obtaining the license, the employee returned to his previous employer. In another case, an employee at the authors' facility was given assistance to prepare for several ASE exams. The employee never took the exams.

Evaluation of Development Programs

A critical component of any development program is the evaluation component. This part of the program must be developed before the delivery of the program and, in some cases, before the content is determined. The facility must know what it intends to accomplish and how it will benefit the company via the delivery of a program. In addition, the evaluation must identify how it will measure the success of the program before encountering any costs. If a manager does not know what is to be accomplished or how it will benefit the company, the program cannot be justified.

The evaluation component also must determine in advance how the content and delivery of the program will be judged. The person or company delivering the program must be made aware of these criteria before agreeing to offer it. For instance, one method to judge the success of the program is to look for evidence of improvement in personnel knowledge, credentials, and performance. The other is to judge the quality of the program delivered based on expectations and/or comparisons with other offerings.

With respect to changes in personnel, the assumption is that if personnel quality, credentials, and performance improves, the company will benefit. For example, a development program's outcomes may be:

- To have all technicians possess an emission license.
- To exhibit that the service consultants can answer all questions posed by customers.
- To create a survey to determine if all customers are highly satisfied with their service.
- To increase the number of daily oil changes.

Another evaluation should be conducted on the quality of the program by examining the content, method of delivery, materials provided, follow-up with employees, quality of instruction, and so on. Most of this information may be gained from the employees who participated in the program. Again, the criteria to be used to judge the program should be shared with the people who will deliver it. The criteria used to determine success must be linked into the program's objectives. Did the evaluation of a program show that the objectives were met?

Final Thoughts

Automobile service facilities are engaged in the business of servicing some of the more complex machines owned by nearly every adult in the country. Next to the person's home, these machines represent the most expensive purchase for the average person. Therefore, when new employees are hired by an automobile service facility, it often has to assume some cost (either directly or indirectly) of getting these employees ready for full-time employment, either as a technician or service consultant. To assist in this endeavor, some automobile manufacturers have created technician-training programs, some of which are located at secondary and postsecondary schools. Even so, automotive facilities must still assume the costs for the induction of new employees into their company, as well as keeping their employees up-to-date. This responsibility will continue because as the automobiles evolve, the people who work on them plus the people who assist the workers with these people must keep up-to-date on the latest innovations.

Consequently, there is little doubt that induction and development programs are important to the operation of a successful automobile service facility. This, of course, requires the development of plans along with the expenditure of time, money, and energy. Their usefulness requires the owners and managers to take the time to determine what is needed, who should be included, how it will be delivered, who will do it, and if the program or person was successful at the completion of it.

Finally, development programs are an important part of the service manager's job because he/she is accountable for the continued success of the facility. A new manager on his/her first day must study the policies and procedures of the facility, and conduct a careful review and assessment of the way the service facility operates. Next, the manager must determine how he/she can improve it. Then, the new manager should set up a development plan to bring about change by training the employees on the way they can improve the effectiveness and efficiency of the facility. A mistake many new managers make is to try to command people to change instantly. They fail to recognize that change is threatening to many people and, as a result, initiating change usually takes time and patience. Therefore, a development program often includes reasons for change and states why it is important to the employees' professional future and personal development.

Review Questions

1. What are the three most important responsibilities of a manager? Explain why they are important.
2. What is the purpose of induction?
3. Describe why induction programs are important.
4. What is the difference between a formal and an informal leader?
5. What is the relevance of *mother duck* imprinting to management?
6. What is the purpose of development and why is it important?
7. What are the two objectives for personnel development programs?
8. List the different types of employee personnel development programs.
9. What are the four subplans to be included in a development program plan?
10. How can ASE certifications play a role in personnel development?
11. What occurs during an orientation program?
12. What are performance standards, targets, and benchmarks?

CHAPTER 8

EVALUATION OF EMPLOYEE PERFORMANCES

LEARNING OBJECTIVES

Upon reading this chapter, students should be able to:

- *Describe the purpose and objectives of employee evaluations.*
- *Define validity, reliability, and objectivity.*
- *Explain the reason for formative and summative evaluations.*
- *Identify sources that can be used to prepare an evaluation of an employee.*
- *Present an evaluation plan.*
- *Describe the shaping process.*
- *Identify the three parts of a performance objective.*
- *Explain how the shaping process and performance objectives are used with a formative evaluation.*
- *Describe the three-step termination process.*

Introduction

The last function in personnel management is evaluation, which is the topic of this chapter. Often referred to as employee appraisal, the evaluation of employee performance is a task assigned to line managers, such as the service manager. Because this often is considered an uncomfortable task, personnel evaluations may be avoided or conducted in haste. This should not be the case. The evaluation of performances should be a constructive experience for the employee and, most of all, it should not be considered a personal assault. Evaluations are simply a part of running a business to ensure the capital invested in human resources is being spent appropriately and efficiently.

Managers are (logically) responsible for the performance of their subordinates. The quality of their work reflects on the reputation of the manager. For example, at one shop, the owner took great pride in the cleanliness of his shop floor. When a technician failed to clean his bay as expected by the owner at the end of each day, the owner took it as a sign of disrespect. The owner never blamed the technician but rather the manager. He believed that it was the manager's responsibility to supervise the employee's work area to ensure that respect was shown for the owner's property. Therefore, because the employees' performances reflect on the perceived competence of the manager, the manager must continually evaluate their work and activities, especially those given a high priority by the owner, to ensure they meet the expectations of the owners and/or upper management.

The manager's evaluation task duties should include both formal performance reviews and informal observations. In the case of the informal observations, strong and weak performances should be documented and maintained by the manager. The formal reviews should include a summation of the informal observations relevant to the review. The formal review with the informal summary should be discussed with the employee, placed in the employee's personnel file, and reported to the owners or upper management. Employees receiving a strong evaluation should be offered a form of positive recognition, such as an increase in pay, whereas those who earned a weak or less than acceptable evaluation should be provided with assistance via a development program (as discussed in Chapter 7). The purpose of this chapter is to describe a businesslike approach to an evaluation process that is acceptable to the employer and respectful of the employee.

The Purpose of Evaluations

In the simplest form, the purpose of employee evaluations is to determine how well each employee performs his/her job. The manager must determine if the employee knows what his/her job requires him/her to do, if the person can perform the tasks and duties required, and the level of an employee's proficiency relative to a job target. To gain the information needed for a report, the evaluation must be considered to be a process and not an event. Specifically, the evaluator must observe the employee over a period of time, such as from one formal evaluation to another, and not just for an hour before the assessment report is due. The manager must document job performances plus observe whether the employee follows the rules and regulations set forth in the operation manual and company policies. A tip from the experiences of the authors is that an informal observation may not seem to be important, such as the failure to comply with company policy, but it may be a hint that the manager should look for other failures to abide by policy as well as errors in the employee's judgment.

Employee evaluations are an important part of the business relationship between an employer and employee. First, an employer provides employees with money in exchange for work; however, the work must provide a financial benefit to the company in return for pay. Second, the employee can expect to have continued employment provided he/she continues to financially benefit the company. Third, the manager must evaluate the performance of the employee to ensure the employer that the employee is providing the financial benefit. When an employee evaluation exhibits satisfactory or exceptional job performance, the employee is typically awarded additional compensation and benefits, such as additional vacation time. If a person's performance is not providing a financial benefit to the employer, then the business relationship is in jeopardy. The employer may offer the employee an opportunity to improve his/her performance (a development program) or terminate the employee. People also may be terminated for violations of certain policies and other unacceptable actions, such as charging the purchase of parts for personal use to the business. If an employee is provided with the opportunity to improve his/her performance, and the performance does not improve after a given period of time, steps toward termination have to be taken.

When termination occurs, it must reference the business relationship; simply stated, the employee does not provide a financial benefit to the employer. The employee should never be told that he/she failed; rather, the termination should maintain the business arrangement to avoid unnecessary distress and angry reactions. Therefore, the approach used in this chapter is that the purpose of evaluation (to determine how well an employee performs his/her job) centers on the business relationship between the employer and employee.

Measuring Employee Performances

When conducting employee evaluations, managers must recognize some basic principles related to the measures of human performance. First, what is being measured must be clear and understood by the evaluator and the person being evaluated. To identify what should be measured, a manager has three major sources of information to use when measuring performance. These sources are:

- Employee job description
- Service unit's operation's manual and
- Company policies

Although there may be other items of interest to a manager when conducting an assessment (such as a specific concern), these three documents are primary sources because the owners/upper management approved them (as explained in previous chapters). These sources provide the manager with all of the tasks, duties, and expectations assigned to an employee. They not only list what the employee is to do, but they also indicate how the person is to conduct him or herself on the job. The manager must then determine whether the employee has performed his duties and conducted himself/herself properly on the job.

In terms of a manager's duties, employee evaluation is one of the most important tasks to be performed for the employer. When an owner hires a line manager (such as a service manager), a major question is whether the manager has the background and expertise needed to evaluate his/her employees. The next concern for an owner is whether the manager can lead them to higher levels of performance. A manager who cannot accurately evaluate performances will very likely not be able to lead subordinates to higher level of performances and, as a result, will not meet the company's goals and objectives. In such a case, the owner's have no choice but to replace the manager as quickly as possible with a person who can accurately evaluate performances and lead employees so that company goals and objectives are met. This means that when an employee is deficient on any measure of performance, the manager must either assist the employee so that he/she can correct the deficiency or find another employee who can do the job.

Consequently, evaluating the performances of a subordinate is an important task. It is difficult to evaluate performances that cannot be measured. As a result, managers should choose a quantitative means to record observations; however, it must be a *valid* method of measurement, which means it must measure the performance or behavior the manager intended to measure. In other words, if the manager intends to measure a job performance of technicians, then he/she must be careful to restrict the assessments to relevant performance measures. For example, a technician's measure of productivity might be the number of flat-rate hours generated or the number of comebacks (second attempts) in a month. An evaluation of a technician's productivity would not be valid if it was on irrelevant or meaningless information, such as the brand of tools the technician owned, or if it was influenced by a manager's personal biases, such as how well the technician's uniform fit. Unfortunately, a manager might not be aware of some of his/her biases. For instance, a technician may approach a job differently than the way the manager (who may have been a technician at one time) would have approached it. This may annoy the manager and because of the differences, the manager may subconsciously have a negative reaction to the employee's performance even though the employee did the job successfully. In such a case, the negative reaction to the performance that is technically correct, but uses a different approach, would not be a valid measure.

Another concern in the measure of performance is *reliability*, which means a measure is consistent from one assessment to another and from one employee to another. For example, a ruler that measures 12 inches will always measure 12 inches, regardless of the time of day, the item being measured, the amount of light and temperature in the room, and so on. In other words, when conducting an assessment, the manager must control as many influencing variables as possible, such as an evaluation conducted in the morning will likely produce a different result than one conducted at the end of a long, hard workday. If the results of the first assessment are different from the second, neither assessment is reliable. If the assessment is not reliable, it cannot be valid.

The problem of reliability also brings up the question of conducting an assessment of a person's *typical performance* versus a *prepared performance*. For instance, assume a manager walks into a bay and, without telling the technician, evaluates the performance of the person when making a repair. This would be an observation of a typical performance. The observations of the manager, however, would likely be quite different if the manager told the technician the day before that he/she would be evaluated the next day at a specific time making the repair. This is a prepared performance because the technician can prepare for the evaluation. This is not to say that a person should or should not be allowed to mentally and possibly physically prepare for an evaluation, but that the manager must recognize that the technician's performance will likely be different when the evaluation is announced versus unannounced. Therefore, evaluations of a typical performance cannot be compared to a prepared performance.

The *objectivity* of a measure also must be taken into account by the manager. This means the subjective judgment of the evaluator must be kept to a minimum. If evaluations can be quantified or at least assigned a numerical value based on the number of acceptable specific performances observed, the objectivity of the evaluation is usually improved. For example, an objective evaluation would document the amount of time taken to complete a job, the number of any errors made (if relevant) when conducting the repair, a check of settings and tolerances, the proper processing of paperwork, the time taken to greet and prepare a customer's request for service, and so on. In these evaluations, the idea is to compare the performance to an expected outcome or standard, which is a **criterion**. When a person's performance is not compared to a standard but to the performance of other workers, then it is considered a **normative** measure. Normative results would indicate if an employee is better than or not as good as another employee who does the same type of work. The weakness in this method is that all of the employees being assessed may not be able to meet a criterion and are performing below an acceptable standard. So a criterion-based method of assessment is recommended, especially for smaller service facilities with few employees.

Two Types of Evaluations: Formative and Summative

Assuming that the method of evaluation is valid, reliable, and objective, a manager may evaluate an employee for one of two reasons. One is to assist the employee improve his/her performance in the conduct of a task or duty. This is called a **formative evaluation.** The formative method is a narrow and focused assessment intended to assist an employee overcome a weakness/problem. In these evaluations, the manager will assess the employee's work performance to determine the cause of the problem or need for improvement. The manager should then assist the employee in finding ways to improve his/her performance through a development program (see Chapter 7). After a set period of time (benchmark), another formative evaluation should be conducted. This practice should continue until the employee's performance is acceptable. Formative evaluations may not be documented

FIGURE 8-1 A formative evaluation is a focused assessment intended to assist an employee in improving performance.

FIGURE 8-2 A summative evaluation is a broad evaluation that covers as many job tasks and duties as possible.

and are often the best tool that a manager can use to improve an employee's, or even a unit's, performance.

A **summative evaluation** is a broad evaluation that covers as many job tasks and duties as possible. These evaluations are formal assessments and, unlike formative evaluations, they are used for personnel decision-making purposes, such as a promotion, pay increase, suspension, or termination. These assessments are documented as shown in Figure 8-2, with at least three copies signed by both the manager and employee. One signed copy is sent to the owner/employer, another signed copy is placed in the employee's personnel file, and the third is given to the employee. Summative evaluations should review an employee's performance at regular intervals (six months, one year, two years, etc.), unlike a formative evaluation, which is conducted when there is a concern about an employee's performance. It is possible, of course, that a new employee will have shorter time spans between summative evaluations (say every three months) and perhaps more formative evaluations than an employee with many years of service. In the case in which an employee has had several formative evaluations and is still unable to perform a task or duty in a job description, a summative evaluation would be conducted to document the deficiency so that formal personnel action can be taken.

Preparing for Evaluations

One of the greatest mistakes a manager can make is to fail to properly prepare for an evaluation of an employee. An evaluation cannot be created from memories of past observed performances, a written statement on the impressions an employee has made during one observation, or materials that are made up the night before the assessments are made. These practices will not be valid, reliable, objective, or meet the expectations of the owners.

The meaning of the title of manager must be taken very seriously because it implies an administrative authority to exercise control over others and a responsibility for their performances. This requires the manager to lead, direct, and evaluate the performances of those employees under his/her control. The criteria used to evaluate the performances of the people being managed must be relevant to the expectations of the owners. The failure to provide relevant and professional evaluation reports to the owners and/or upper management based on proper criteria could likely be a career ender for a manager.

The criteria to be used for the evaluation of an employee must be collected and reviewed for each individual. Although some criteria (such as the proper processing of paperwork) may apply to several employees, other criteria would be specific to each individual position. The sources for the information needed to identify the criteria for an evaluation are contained in the:

1. Employee's job description
2. Operation manual for the unit
3. Company policies
4. Employee's personnel folder

From these sources of information, a manager must select the tasks and duties in the job description to be assessed.

Using the operations manual, a list should be prepared of the rules and regulations that are relevant to the employee's position, such as a dress code that requires hair to be short or tied back so it does not pose a safety problem. Next, any company policies that are not in the operation manual but need to be emphasized or reinforced should be checked, such as the maintenance of a drug-free workplace. For instance, an employee may not use drugs but needs to know that the manager is checking and reports are prepared on their use. Finally, the employee's personnel file must be reviewed. The review should pay particular attention to previous evaluation reports, the induction program review, and any warnings or reprimands.

After the review of documents, the performances to be evaluated must be selected. Next, the manager must set standards for each item that will become the criteria to be included in the evaluation. The standards must indicate what the manager expects to see or not see in terms of the employee's performance. Finally, the manager must decide how to score the observations. For instance, a score may offer two choices: acceptable or not acceptable. Others may allow a mark from 1 to 5 (1 = poor, 2 = below average, 3 = average, 4 = above average, and 5 = exceptional). A third could permit the manager to enter a written comment about the employee's performance.

After the evaluation materials are prepared, the manager must decide how to collect, record, summarize, and reach a conclusion based on the performances observed. Obviously, as discussed previously, one source of information would be from the notes taken by the manager since the last evaluation. Others would include:

- Observations of the employee since the last performance review
- Records or samples of completed work
- Interviews with the service consultant, parts specialist, staff managers, technicians, and even customers
- An interview with the employee who is being evaluated
- Customer surveys

Before the evaluation of an employee is conducted, all of the materials and procedures must be checked for validity, reliability, and objectivity. A suggestion is to conduct a field test by collecting information, preparing a report, and presenting it to a colleague, such as the parts department manager.

The Evaluation Plan

After the evaluation materials have been prepared, the manager must prepare a plan to conduct it. Usually the evaluation of an employee is conducted on the anniversary date of their being hired or given a permanent position in the company. Regardless, because it takes considerable time to perform an evaluation, it is a good idea if a manager puts employees on a rotation so that too many are not conducted at the same time. In the plan, the first step would be to meet with the employee and explain what will be used to examine his or her performance and the way it will be recorded. In addition, at this meeting another meeting with the employee should be scheduled. This meeting is the last step in the plan and is set to discuss the report before it is forwarded to the upper management and/or owners. From the experience of the authors, this first meeting sets the tone that the evaluation is a formal and serious process based on a business relationship between the employee and employer.

The second step in the plan should present the activities of the manager in the collection of the evaluation material. For example, the informal observations of typical performance and behavior should be an ongoing activity from the last evaluation to the present. The dates to conduct formal observations, to interview other employees, to collect work samples, to review the person's personnel file, and so on must be placed in the plan and on the manager's calendar. Again, from personal experience, the authors recommend these dates be placed on the manager's calendar; otherwise, they are put off or delayed indefinitely. When this occurs, the seriousness of the evaluation to the employee will be lost.

Third, the plan must clearly indicate how the performances will be scored, summarized, interpreted, and reported. As noted above, this part of the plan should ensure that the assessment is as objective as possible. It must be clearly explained to the employee in the meeting in step one. It cannot be made up as the assessments take place or after the information is collected. When this occurs, the validity of the evaluation is compromised because the manager will be suspected of measuring performances that are his/her personal preferences that will give some employees a break while being tough on others.

All summative evaluations, including the one at the end of the induction period, must include a recommendation from the manager in the report to the owners or upper management. For example, the report may recommend that the employee be retained and given a small, medium, or large increase in pay, be retained with no increase in pay, or not be retained. When an evaluation recommends that a person not be retained, or even when an increase in pay is not recommended, a second or even third manager may be asked to collaborate on the evaluation. At smaller service facilities, this may not be possible; therefore, an additional a person (such as a senior employee or owner) may be asked to make observations of performances, interview an employee about the behavior of the person being evaluated, or interview the employee being evaluated. The inclusion of the additional managers or others improves the validity of the evaluation and reputation of the manager.

The last step is to review the results of the assessment with the employee. The employee should have a copy of all material with the exception of confidential statements taken from other employees or customers and miscellaneous references. The manager must take the time to go over the findings and conclusion in detail. The employee should have time to react or respond as the material is reviewed. Finally, the recommendation of the manager should be read to the employee and a line must be provided at the end of the report for the employee to sign, indicating the information was reviewed and understood. The signiture does not mean that the employee agrees with the report. In some cases, the company policy or a union contract provides for the employee to appeal an evaluation report. After the manager and employee sign the report, a diagonal line should be drawn from the signatures to the bottom of the page. This is to prevent anyone from inserting any postscript comments and declaring them to be a part of the report.

Problem Personnel, Performance Objectives, and Formative Evaluations

As noted above, formative evaluations assist employees with performance problems. The intention is to help an employee as opposed to terminating him/her after receiving a poor summative evaluation. Because time and money has been spent to hire the person, the investment in formative evaluations is justified. In addition, managers should continually work with employees to improve their performances and, therefore, the performance of the facility.

Development programs (see Chapter 7) for employees should use formative evaluations in the **shaping process**, which has been used by educators for many years. Basically, the shaping of employees is conducted by reinforcing acceptable performance and redirecting performance deficiencies.

Therefore, the focus for shaping employees is on performance. Consequently, performance objectives identify the behavior to be reinforced plus a target to determine when the performance is acceptable.

Writing Performance Objectives for the Shaping Process

A performance objective is a statement that is typically one sentence long and contains three basic elements: a condition, the performance, and target. The sentence starts with the condition and in a few words tells under what conditions the employee will demonstrate a performance. For example, a performance objective sentence might start with the phrase "When changing oil . . ."

Next, the sentence identifies the performance to be assessed. There are three types of performances an employee could be expected to demonstrate; however, a performance objective should include only one of three performances. These performances are:

1. Knowing
2. Doing
3. Behavior

Finally, the performance objective statement ends with a target to assess the employee's performance. Using the above performances, an illustration of an appropriate target for each might be as follows:

Condition	Target
Knowing	A technician must identify where to find the specifications for automobiles.
Doing	The technician must perform a job duty or procedure, such as an alignment.
Behavior	When a service consultant has an angry customer, the consultant must act in a professional manner.

In other words, when the target is met, the manager knows that the employee has met the performance objective's expectation. The purpose of the shaping process is to direct the employee toward the target by reinforcing appropriate behavior (such as compliments for a proper performance) and redirecting deficiencies (providing additional instruction or direction). When the person is ready, a summative evaluation may be conducted to ensure the employee can meet the assigned objective.

Illustration of the Use of the Shaping Process

Assume that a manager observed that Elton, a technician, had pools of oil on the floor of his bay throughout the workday. Further observations revealed that when Elton changed oil, he was not using the waste oil device properly. Because the oil on the floor created a safety hazard, the manager asked Elton why he did not use the waste oil device properly. If Elton did not have a reason, the manager could prepare a performance objective to guide him for the remainder of the week. The performance objective would state, "When doing an oil change (the condition), you will properly use the waste oil device (the performance) and the floor will not have any oil on it when the oil change is completed (the target)." The manager, or perhaps an area supervisor, should write the performance objective on his notepad and give a copy to Elton to hang on his toolbox, put in his shirt pocket, or to attach his clipboard.

Elton would then be given instruction on how to properly use the waste oil drain device. Then, if the manager or area supervisor observed Elton using the equipment properly and there was no oil on the floor, he would be complimented. A summary evaluation could be conducted at the end of the week and, if successful, a complimentary letter could be given to Elton with a copy placed in his file. If the manager or area supervisor found that Elton's bay continued to have oil on the floor during the week and observed that he was not using the oil drain device as prescribed, the manager must determine the reason for the failure, such as:

- A need for additional training (given additional instruction and practice)
- Defective equipment (a new or better oil drain device was needed)
- A refusal to use the device properly

If the manager found that Elton had a pool of oil on the floor because he refused to use the oil drain device, as directed and shown in Figure 8-3, then the manager would have to present him with a written reprimand for insubordination. A charge of insubordination or the inability to perform work properly would be justified because he was given instruction and then observed doing it incorrectly. The charge of insubordination should not permit an employee to charge the employer's account for unemployment in the event that termination would result; however, a charge of an inability to perform a duty likely would.

FIGURE 8-3 Oil drain device.

Use of Performance Objectives in Development Programs for Company or Unit Purposes

In the last chapter, it was explained how personnel development programs can be used when a company or a unit in a company has any changes in its mission, goals, objectives, or procedures (strategic financial plan), as well as changes in leadership, strategies, or operations. These personnel development programs also may use performance objectives to guide employees. Again, the performance objective statement for the group would contain a condition, performance, and target to be attained.

For example, assume a service consultant in a facility that services a fleet of trucks was told to inspect the brakes each time they were serviced. To implement this inspection, he could create the following performance objective for technicians. "When fleet trucks are serviced (condition), the technician will inspect the brakes (performance), document the results, and recommend repairs when they do not meet the state's inspection guidelines for safety (target)." The service consultant could review the technician's performance after each oil change by checking for the documentation on brake inspections (target) or ask the team leader to see if they are entered. Of course, in this delegation to a team leader, the service consultant is still responsible for ensuring that each truck meets the state safety specifications. As explained in Chapter 1, when job responsibilities are assigned to a position, they cannot be delegated. The service consultant may assign a responsibility to a team leader, but the service consultant cannot transfer his/her personal responsibility to upper management or the owners.

Termination

With respect to having to deal with problem personnel, the recommendation is that managers should always assume that the employee wants to comply with company policy and procedures. They must believe that all employees want to do a good job and to meet the standards set for performance objectives. When this does not occur, the manager should first use the shaping process and performance objectives, as discussed earlier. If an employee still does not or cannot meet the performance standards, the manager must conduct a summary evaluation and properly document the evidence that the standards were not or could not be met (or improper conduct continued).

The termination process begins with a written warning to the employee that contains: (1) the problem; (2) the documented evidence of the problem; and (3) a warning that if improvement is not made, termination will be the consequence. The written warning should be handed to the employee by the manager (not placed at his/her desk or workstation or mailed to his/her home) and then read to the employee by the manager. The manager should document the meeting and the reading of the letter on a note attached to the copy placed in the personnel file.

The difference between a two- or three-step process regards the next action to be taken if improvement is not made after the first written notice. In the two-step process, if improvement is not made, the person is terminated or suspended. In a three-step process, a second written letter is delivered and the same procedures described earlier for the first letter are followed. If improvement is still not made, the person is terminated or suspended.

A suspension period may be with or without pay. A *with pay* suspension is usually when an investigation period is justified, such as a case in which financial accounts must be reviewed to determine whether the employee was negligent or if fraud was committed and termination is required. When a person is terminated, the manager should deliver the letter to the employee, read it, have the employee immediately remove all personal belongings from the building, hand in keys and other company belongings, receive a final paycheck, and leave the building. If the employee has possession of any company belongings, such as uniforms, to be returned, the final paycheck must be retained until the company belongings are brought to the manager. The removal of personal belongings may be done on a weekend or after the business has closed for the day or evening. This is simply to relieve the terminated employee of unnecessary embarrassment.

When a manager knows a termination will take place, it must be conducted with other employees in mind. Even if the behavior of the manager and employee is respectful and businesslike, terminations are disruptive to the workplace. Consequently, the authors recommend that terminations take place on Friday afternoon at the end of the day (unless circumstances dictate that the employee must leave the premises immediately) and, if necessary, personal belongings (such as tool boxes) be removed under the manager's supervision after working hours or on the weekend. The final parting words by the manager to the person should not be hostile but indicate regret that the person was not able to perform at the level expected, that he/she was not a failure, and to wish them success in the future.

Recognitions

After a successful summary evaluation, an employee usually receives some form of recognition, such as an increase in pay. Recognitions also should be given when an employee meets a performance standard, completes a development program, or does something that benefits the business. The failure to recognize accomplishments, either formally or informally, can sometimes do more damage than a reprimand that is not deserved. Such failures often indicate a manager is not paying attention, does not care, or does not appreciate the effort. Employees, therefore, often have an unfeeling or callous attitude toward the manager and company.

Recognitions for extra work, as seen in Figure 8-4, may be as simple as saying, "Thanks for staying to finish Mr. Carlucci's car. It really helps the business when we exceed the customer's expectations. We appreciate

FIGURE 8-4 When a technician stays late to finish a customer's car, the effort should be recognized.

what you did and hopefully, we can return the favor." Others may be more formal, such as an award at a recognition dinner. In all cases, a manager should attempt to give privileged treatment to those employees that give their personal support to the company, such as the use of a bay to work on their car after business hours. If an employee is not supportive, they should not receive the same privilege. Rather, these employees should be told that when they support the company, the company will go out of its way to support them. By support, the idea is to support the business operations of the company via performance and not the offering of fake or phony personal expressions of admiration of the manager, upper management, owners, and so on; claims of hard work; declarations of a positive work attitude and love of job; confidential reports about the failures of other employees; or looking busy but not accomplishing anything. Results, not words, must be shown.

Final Thoughts

Employee evaluations cannot be presented when managers can only make undocumented claims. When this occurs, meetings with employees usually result in claims and counterclaims. When these arguments are in front of the owners or a labor review board, everyone loses, especially the manager. In the authors' experiences, when failures to perform a job or charges of improper conduct are documented and evidence is presented in a businesslike manner, there is hardly ever an emotional confrontation. The exception, in their experience, is when an employee has improper or illogical perceptions and believes his/her action was justified or done correctly. In one case, for example, an employee failed to report to work and in another case, an employee refused to follow orders. Both believed their actions were justified even though both caused considerable problems for the business. Their actions could have been dismissed with a warning, but both truly believed they had not done anything wrong. Both were terminated.

The evaluation process also may be impacted by a collective bargaining (union) contract. When employees of a company are working under a collective bargaining contract, the process is likely included or at least mentioned in the contract negotiated between the owners and union leaders. Managers who are responsible for the conduct of the evaluations must read the contract to be sure they do not violate any agreements in it. Most important, when a manager meets with a union member regarding an assessment of his/her performance, the union representative (shop steward) should be present. Abiding by the contract and working with the union representative to get the job done, which in this case is the evaluation of employee performance, is smart business. For example, a contract will spell out agreements between labor and management such as pay rates, days off, number of hours worked per week (as per labor law, discussed in Chapter 5), the pay method (hourly, flat rate, flat rate with a guarantee), and how employees are written up among other key issues that have to do with employees and working conditions. Items not included in a bargaining agreement are labor rates charged to customers, hours of service facility operation, maximum amount of work that can be performed by a technician, and the type of work assigned to a technician among other management related prerogatives. Because the evaluation of employee performance is important to a union, they often attempt to negotiate conditions on evaluations, suspensions, and terminations. The employer, however, tries to implement an evaluation process similar to the one described earlier, but with little or no interference from the contract.

Because the evaluation of an employee's performances and behavior creates anxiety, the approach and attitude of the evaluator is important. If it is handled in a clumsy fashion and if the attitude of the manager is uncertain or negative, the process will be more threatening than necessary. In addition, some people are terribly sensitive about receiving and giving critical reviews; however, it is a part of a manager's responsibility. A manager will be perceived as good as his/her workers and a worker is only as good as his/her best critic. Therefore, if evaluations are not conducted or are conducted poorly, everyone (manager, employee, and owner) loses.

Finally, as the authors have found, likely the most difficult situation is when a person just cannot get the job done. The person may try hard, have a great attitude, ask for assistance, and still not be able to get the job done. For example, as the authors have learned, some people want to be an auto technician and for some reason or other, they just cannot do the work. For the manager, this is difficult; however, the greatest favor the manager can do for such a worker is to counsel him/her into another occupation. A person will not be happy if he/she continues to try to do something he/she cannot do successfully. Instead, people must find what they can do well and for which they will be rewarded. Sometimes the manager has to help his/her people to find out what that job may be, either with the company he/she is working for or at another company.

Review Questions

1. What are the purpose and objectives of employee evaluations?
2. Define validity, reliability, and objectivity.
3. What is the reason for formative and summative evaluations?
4. What sources can be used to prepare an evaluation of an employee?
5. Describe the elements of an evaluation plan.
6. What is the shaping process?
7. What are the three parts of a performance objective?
8. How are the shaping process and performance objectives used with a formative evaluation?
9. What is the three-step termination process?

PART II

PRACTICAL EXERCISE

Small Group Breakout Exercises

You are the Service Manager at an automobile service facility/ dealership. The business is an S Corporation. Your services include general diagnostic work, repairs, and maintenance on all models.

You report to the two owners of the corporation. One owner is the Corporate Executive Officer (CEO) who also performs the duties of a general manager in charge of service operations. The other owner is the Chief Financial Officer (CFO) who also performs the duties of the business manager. They share an administrative secretary.

As the service manager, you have two people reporting directly to you: the Service Consultant and the Parts Specialist (at this dealership, the CEO has required that the Parts Specialist reports to you even though his office is in the parts department). In addition, there are five technicians employed in your department: one A-tech, two B-techs, and two C-techs.

In terms of the facility, there are five-fully equipped service bays with a workbench, vise, lift, compressed-air hose, and access to water. In addition there is a room devoted to the storage of other shop equipment such as jacks, stands, and special tools. There is also an oversized work area that has tire-mounting and -balancing equipment, a brake lathe, and a solvent tank that sits next to a service bay contains an alignment machine.

Management that believes the dealership facility can provide a customer with a wide range of services. In addition, they feel that they meet the automobile manufacturer's franchise agreement requirements with the exception of towing and major body repairs, which must be subcontracted to outside repair facilities.

To assist the CFO, there is a bookkeeper who supervises one full-time and one part-time cashier. In addition, the dealership employs a maintenance supervisor who handles facility repairs as well as two part-time employees responsible for keeping the building clean and grounds maintained.

1. In order to have a thorough knowledge of the different jobs that may need to be filled as the dealership expands, the CEO and CFO want a job description for all employees who are under your management. This includes descriptions for an A-tech, B-tech, and C-tech position as well as the service consultant and parts specialist. (Note: the descriptions must correspond to your answer to Question 2 in Part I Practical Exercise.) As part of your answer, you must incorporate the following information into your job descriptions, as appropriate for the position.
   A. As part of the manufacturer's franchise agreement with the dealership owners, the technicians must each complete a battery of training courses so that your department can be paid under the manufacturer's warranty. You must ensure that each technician's job description contains the appropriate wording. The following are the manufacturer training requirements for the types of repairs performed by the dealership:
      i. Electrical Repairs: three training courses.
      ii. Automatic transmission repairs: the electrical repair training courses plus four transmission-specific training courses.
      iii. Air-conditioning repairs: the electrical repair training courses plus one A/C-specific training course.
      iv. Brake and antilock brake system repairs: the electrical repair training courses plus five specific brake system courses.
      v. Engine performance and engine repairs: Both the electrical repair training courses and the air-conditioning repair training course plus two engine performance training courses and one engine repair course.
      vi. In addition, each dealership technician must complete the new-model training course prior to the release of the next model year vehicles otherwise the technician's ability to perform warranty repairs will be suspended until completed.
      vii. Furthermore, your state has an emissions program that requires technicians, who perform vehicle emissions inspections, to pass a course and recertify every two years. You must include this in the appropriate job descriptions.

2. Your A-level technician quit today. Provide the CEO with a plan to recruit and select a replacement.
3. After screening the applicants for the A-level technician position, you must prepare for the interview process. The CEO plans to participate in the interviews. Prepare an outline for the process, the questions you and the CEO will ask each candidate, and the method you will use to rank them to make the hiring decision.
4. The new A-level technician was hired and must be inducted into the department and his new position. Prepare an outline for an induction plan that indicates what the new technician will do in the first week of his/her employment at the dealership plus a long-term induction plan that recognizes the following:
   A. The new technician came from another state and does not have the required state emissions license. The test will be given next month and the next emissions course will be given in three months. It takes a month to process all paperwork after taking the test. He/she will not be able to conduct any inspections until he/she receives formal notification he/she passed the test. The cost of the test is $200.
   B. The new technician also needs to take the manufacturer's electrical course, an automatic transmission course, and a new-model training course to be eligible to perform warranty repairs.
      i. The new-model training course is given online and can be taken at any time. The course takes approximately one day.
      ii. The transmission and electrical courses must be taken at the training center. Both are given next month, but on the same day as the state emissions course, and then again in three months, again on the same day as the emissions course. Therefore, the technician can take only one course at a time. The cost of travel to the center is approximately $300. If he/she takes the electrical course, he/she could take the automatic transmission course in two months in the next district, but this will cost $1,100 to cover a plane ticket and hotel accommodations. The dealership cannot perform automatic transmission repairs until the new technician completes the course (approximately three automatic transmission repairs are done each month; each with a profit of around $125).
      iii. In your induction plan for the new employee, include the additional training needed with costs (note: the new employee is paid $20 per hour) for the owners to review.
      iv. Based on the certifications the new employee lacks, indicate the types of jobs he/she will not be able to perform over the first six months of his/her employment and how that will affect your review of the employee's performance.
5. Your technicians must be kept up-to-date, and it also is the owner's wish for the B techs and C techs to be advanced to the next level of competence. Create a development plan based on information found in Question 1 for the A, B, and C technicians, including provisions for the advancement of the B and C techs to the next level.

Consult the appropriate chapters in this part for examples of the type of work and A-, B-, and C-level technicians are expected to do.

6. As part of your job, you must evaluate the Service Consultant and Part Specialist. Prepare an evaluation plan for their review. Support the rationale for your evaluation with citations from the appropriate textbook chapters.

7. In your evaluation of the Parts Specialist, you discovered several serious performance problems. Although he has a thorough knowledge of parts, your review indicates that he is not able to perform his duties properly. More specifically, his performance failures included delays in the delivery of parts to the technicians as well as errors made when parts are ordered. In these cases, the time for receiving parts or reordered parts added to the time the technicians took to make repairs and caused a loss in revenue. For a meeting with the CEO and CFO, you are to prepare a plan that explains how you will handle the unsatisfactory performances of the Parts Specialist.

Prepare a schedule of reviews with follow-up meetings with the Parts Specialist in the days to come. With your schedule, include a description on how you will attempt to improve his performance, and conduct reviews and meetings allowing for the possibility of termination.

8. Friday afternoon, 2/12, you had to meet with the CFO and manufacturer's representative at the country club. Today is Monday 2/15. It is 7:30 A.M., and there are a number of situations in your office that you must handle. First, begin by reviewing the list and then place each situation in the order you must take care of them. Second, plan your day by blocking out time to take care of each situation (the importance of the situation will dictate whether it must be taken care of today, whether the situation can be delegated to another employee, or whether a meeting must occur with the CFO/CEO or other hired professional such as an attorney). Third, state the action(s) you would take, assuming you would take any action, for each situation. Fourth, give the reason(s) for your action. Fifth, present your observations about the reason or cause(s) of these situations and identify any patterns that point to a larger problem. When it is apparent that a larger problem is present, state how you would take care of this larger problem, for example, with:

- A review of the work environment
- Revised rules and procedures
- An examination of the systems
- Conduct performance reviews
- A plan of personnel actions (such as additional development or training, verbal warnings or counseling, written reprimand, suspension or termination)
- A change in your own management behavior

9. On your desk Monday morning, 2/15, is the following:

A. A memo dated Thursday 2/11 4:30 P.M. from the service consultant handling the reservation for the A-technicians Automatic Transmission training course in the next district. The hotels are all booked up because of a convention. There are no rooms available in any of the other hotels and motels in the area. If you respond by Friday 2/12, the hotel manager may be able to reserve a suite but the cost is $300 more. This memo was attached to the service consultant's copy of the repair orders submitted to you at the close of business on Friday 2/12.

B. A letter dated 2/6 from a customer's attorney directing the dealership to respond by Friday 2/12 about a faulty brake master cylinder the dealership installed on 1/15 that caused the customer to be in an accident. On 2/12, the attorney planned to file a complaint with the State Attorney General with a lawsuit against the dealership for damages. The letter claims the service consultant did not return previous phone calls made in January about this matter, so the attorney has no recourse but to file the appropriate legal documents. This letter was attached to the service consultant's copy of the repair orders submitted to you at the close of business on Friday 2/12. The registered letter was signed for and opened by the service consultant on 2/9.

C. A reminder note to yourself written during the afternoon golf game at the country club on Friday 2/12 with the manufacturer's representative and the CFO. The manufacturer's representative told you that he personally sent a letter dated 1/9 to the parts specialist that notified him of a recall of defective master cylinders. The manufacturer's representative followed up with a phone call to the parts specialist on 1/14. He asked why the dealership had not sent back any defective master cylinders to date.

D. An e-mail on your computer received on Friday 2/12 at 4:30 P.M. from the parts specialist reporting that the manufacturer's representative notified him on 1/9 about a shipment of defective master cylinders sent to the dealership. He explained that the dealership sold three of them to customers and wants to know what he should do. He further explained that he was worried about his review earlier in the week because it was not very good. He promises to do better.

E. On your phone voice mail is a message from the CEO received on Friday 2/12 at 1:30 P.M. The CEO said that some technicians stopped by his office to inform him that the new A technician was part of the Teamsters Union at his last job. The new technician liked being represented by the union and thought the technicians at the dealership should meet and discuss how to organize themselves to join one. The technicians asked the CEO if they could leave work early to go to the Mangy Moose Bar and talk about it. They assumed that you, the service manager, wouldn't mind as they couldn't find you. The CEO wanted you to call him immediately to discuss their request to leave work early and the union concept in general. (Remember, you were golfing at the country club and didn't get the message until Monday 2/15.) The service consultant (overhearing you listening to the voice-mail message on speakerphone) informs you that all of the technicians left work an hour early on Friday.

F. A written phone message from the service consultant dated Wednesday 2/10 at 3:00 P.M. "The Community College Automotive Technology Department called to remind you about your management presentation Monday afternoon 2/15 from 3 to 4 P.M. They need the title of your presentation by Friday for their newsletter and the weekend newspaper advertisement. They also wanted to remind you that the semi-annual automotive advisory board meeting would be at 4:30 and as want your input on new equipment the program should consider purchasing." This message was attached to the service consultant's copy of the repair orders submitted to you at the close of business on Friday 2/12.

G. A reminder note to yourself that, when playing golf on 2/12, the CFO stressed that he wanted a report on the job descriptions (details found in Question 1) on his desk Tuesday morning, 2/16. He explained he had to review it and discuss it over golf with the bank president on Wednesday morning 2/17. Over the weekend, the CFO left a voice-mail message requesting that you deliver the report today by 4 P.M. You worked on it over the weekend but have another four hours of work to get it done.

H. Your B technician left a note on your desk on Friday afternoon, 2/12, saying that his wife, who is the administrative secretary to the CEO and CFO, was not out sick on Friday. Rather, she called in sick so she could be with her boyfriend, who happens to be a car salesman at the dealership.

I. A note dated Saturday 2/13 from an unnamed technician that informs you that the B technician in letter H above told the other technicians at the Mangy Moose that he planned to run over his wife and the salesman when the salesman comes to work on Monday morning at 9 A.M.

J. Friday afternoon's (2/12) payroll and other bills that were paid by check on Friday were dependent on a Monday morning direct deposit payment from an aftermarket warranty company that owes the dealership $8,900. A voice mail from the aftermarket warranty company on Saturday (2/13) at 1 P.M. informs you that they will not release payment on Monday morning because the overnight package of warranty documentation was not received. On searching the office, you found that the package was still sitting on the service consultant's desk. He had forgotten to call the overnight express mail company for pickup on Friday afternoon after the documents had been copied. Approximately 35 different checks written by the dealership could bounce (at a cost of $25 dollars each) unless $8,900 is deposited at the bank on opening at 9 A.M. Before you left to play golf, you told the service consultant to list the deposit of $8,900 as received on the Friday report given to the CFO. It is 8:00 A.M. and the CFO will not be in until 10:30 A.M. He does not wake up until 9 A.M.

K. A note from the receptionist on Saturday says, "The country club looked for the club you lost on Friday and found a pitching wedge on the 17th hole. It is at the pro shop for you to examine." You can only hope that is it the same one that the CEO loaned to you on Friday. The CEO plans to go golfing Monday after lunch and would probably appreciate having his pitching wedge back.

See what happens when you go golfing?

PART III

FINANCE

CHAPTER 9

BASIC BUSINESS PRACTICE

LEARNING OBJECTIVES

Upon reading this chapter, students should be able to:

- *Describe a transaction.*
- *Explain the purpose of accounting.*
- *Describe measures of income, revenue, expenses, and expenditures.*
- *Given the income and expenses for a period, calculate the profit earned.*
- *Explain the purpose of inventory control.*
- *Outline internal control procedures for receipts and expenditures.*

Introduction

This part of the book covers the terms, reports, and statements that service managers must use to conduct a financial analysis for management purposes. It does not attempt to teach accounting but, rather, how to use accounting information. To open Part III, this chapter starts with definitions of selected terms and the meaning of a transaction. Then, because records of transactions are made of the money received from sales and the money spent to run a business, the chapter explains how this information is used to compute the net profit of the company.

Ensuring that a company's net profit benefits from all receipts and expenses is a concern for all businesses. This is a major responsibility of service managers. They must make certain that money received and spent by the facility is properly handled, recorded, and reported. They must carefully monitor the receipt of money from sales and the use of it to generate sales. This oversight begins with a set of rules and procedures to be followed, which are referred to as *internal controls*. When receipts and spending are protected from abuse and correctly reported, the measures of business activities, such as sales, cost of sales and expenses, are valid. If they are not valid, then they cannot be managed. In other words, decisions made with incorrect information are not relevant. Assuming that proper internal controls are in place, the remaining chapters in the book discuss financial reports, their analysis, and the management of company resources.

Basic Business Related Terms

The definitions of several accounting and business terms used in this section of the book are as follows:

Accounting:

A process that records, classifies, and summarizes business transactions.

Benchmark:

A target to be reached or met as of a specific date or end of a period of time, such as a week, month, quarter, or fiscal year.

Business transaction:

A business transaction occurs when money or something of value is exchanged for goods or services.

Costs:

Selected expenses incurred for the delivery of a specific service or in the sale of a specific product. For example, the cost of running a quick lube and oil bay would be determined by adding up the expenses incurred to run the service.

Enterprise resource planning system (ERP):

A system that uses information in traditional accounting statements and reports, as well as nontraditional accounting information, to make management decisions.

Expenditure:

The spending of revenue awarded to a business or government body.

Expense:

The spending of income earned by a business.

Financial accounting statements:

Reports that present the financial condition of a business and used for tax purposes as well as managerial decision making. The most common statements are called the income statement, balance sheet, and cash flow statement.

Income:

Money earned from the sale of goods and services.

Nontraditional accounting information:

Quantitative information, not typically found in traditional accounting statements and used by managers to gain insights into business operations to make management decisions. Nontraditional measures in an automotive serice facility may include number of comebacks, available technician flat-rate hours, technician flat-rate hours produced, clock hours, number of flat-rate hours sold per vehicle, and customer satisfaction index among others. Nontraditional information is used in the enterprise resource planning system (ERP).

Profit:

When income exceeds expenses.

Qualitative information:

Information about quality from observations, insights, judgments, and perceptions, such as customer opinions about the quality of repairs and services. Qualitative information is subjective and assists managers to gain insights into business operations and is used by the enterprise resource planning system (ERP).

Quantitative information:

Information gained from measurements.

Revenue:

Money received from a government body for goods or services to be delivered and from investments, such as interest earned or dividend paid. In some textbooks, the term revenue is used interchangeably with the term income.

Target:

A proposed objective to be obtained at the end of a specific time period. Targets may be quantitative measures, such as profit, sales, expenses, or number of customers served; or qualitative measures, such as subjective judgments about customer satisfaction based on opinion surveys. For this part of the book, targets would be based on traditional and nontraditional accounting data.

Business Transactions

An accounting entry is made every time there is a business transaction, which is when an economic exchange occurs. This requires that money or something of value is exchanged for goods or service. A service facility enters into numerous business transactions every day.

When transactions bring in money from customers from the sale of goods or services, it is a **sales transaction**. When a facility pays money for good or services, it is an **expense transaction**. Transactions that bring in money from sources, such as interest received from loaning money to a bank through a savings or checking account, is referred to as a **revenue transaction**. When money is received from the sale of goods or services less any sales returns, the total amount is referred to as **gross sales.** When earned revenue from the beginning to the end of a business period (day, week, month, year) is added to the gross sales, the amount is referred to as a company's **gross income**.

After customers are served, they are charged money for the service and goods purchased. The money collected minus the amount of any sales tax paid is the gross-sale amount. All of the money collected from the customers in cash, check, or credit card payment less the sales tax pays salaries of the technicians, parts for the customers' cars, and the expenses to run the business. As shown in the example in Figure 9-1, after the technicians and parts are paid from the gross sales the remaining balance is called the **gross profit**. The gross profit is then used to pay the salaries of the other personnel, including the salaries of the managers and service consultant, plus the overhead expenses. Overhead expenses include rent, heat, light, telephone, uniforms, insurance, and the other expenses incurred when running a service facility. The balance of money left over after the expenses are deducted from the gross profit is called the **net profit.**

The owner must pay the taxes of the service facility from the net profit. The balance is the profit after taxes and is the amount the owner may claim for a return on the investment of money and time spent working in the service facility. When the expenses cannot be paid by the gross profit, the negative balance is called the **net loss**. The money needed to cover a loss must come from the owner. If losses persist, then the service facility will be bankrupt and must close.

GROSS LABOR and PARTS SALES
- <LESS> cost of the technician's labor
- <LESS> cost of parts to service the customer's automobile

EQUALS: GROSS PROFIT
- <LESS> overhead expenses (Examples)
  - Managers' salaries
  - Service consultants' wages
  - Rent (includes real estate taxes)
  - Heat
  - Electric
  - Insurance and employee benefits
  - Other expenses

EQUALS: NET PROFIT or LOSS
- <Less> Taxes on the service facility's profit

Equals: OWNER'S INCOME

EXAMPLE OF FORMAT PRESENTING MONTHLY INCOME AND EXPENSES

GROSS SALES		
Labor	$30,000	
Parts	30,000	
Total Gross Sales		$60,000
COST OF LABOR AND PARTS		
Technician's Labor	$16,000	
Parts	20,000	
Total Cost of Labor and Parts		36,000
GROSS PROFIT		$24,000
EXPENSES		
Management Salaries	$ 5,800	
Service Consultants	4,300	
Rent	3,600	
Heat (oil)	700	
Electric	400	
Insurance and Benefits	1,500	
Other Expenses	1,700	
Total Expenses		18,000
NET PROFIT		$ 6,000
Business Taxes		1,200
OWNER'S INCOME		$ 4,800

FIGURE 9-1 Report of gross sales and profit.

Internal Control of Income

An important responsibility in a facility is to protect the customer's receipts. When payment is made by cash, checks, or wire transfers, procedures must be established so that the money is secure from the time it is received to the time it is deposited in a bank. The procedures designed to protect receipts, as well as the employees who handle the money, are referred to as *internal controls,* which should be implemented in response to the company policy. The internal control procedures should designate the position or positions (not employee or employees) that may receive money, require that receipts be given to customers, designate another position (not person) to make bank deposits, and require transmittal forms be filled out when money is passed from one employee to another. In addition, deadlines must be established for the length of time money may be held until deposits are made. Regulations also must specify the minimum/maximum amount of cash that may be on site at the service facility at any given time. The minimum amount may be held on hand at the end of the day to be available when the business opens. When the amount of cash exceeds the maximum amount, a bank deposit must be made.

The internal controls should direct how deposits are to be made and recorded. A form used for the deposit of receipts is shown in Figure 9-2. The information needed to make the entries into the accounting system is provided on this form plus nontraditional accounting information (number of LOFs and repair orders plus the invoice numbers and amounts) is also presented. Using this form, an internal audit can easily and quickly be conducted. The nontraditional accounting information can be used to monitor the levels of business activity and for comparisons to the traditional accounting information. In other words, exceptions can easily be noted and checked by a manager, accountant, or owner.

The basic rules for handling receipts in a business should involve at least four positions as follows:

1. A service consultant or cashier (a cashier station is shown in Figure 9-3) receives payments for customers and marks invoices paid (when cash and checks are received by mail, two employees should open, record, and transfer the payment).
2. A service manager (or owner) deposits the money at the end of the day (using a drop box at a bank).
3. An accounts receivable clerk, bookkeeper, or owner records the daily deposits and credit card transactions into the accounting records.
4. An owner or treasurer reconciles the checking account.

In the case of wire transfers (a deposit from one bank to another via electronic transfer), the transmittal and deposit should be verified by one employee, entered into the accounting records by a second person, and confirmed in the bank reconciliation by a third person. Note that the

RENRAG AUTO REPAIR, INC.

RECEIPTS FOR THE DAY

Date: ___________

Cash:		Checks:		Credit Card	
IN#	Payment	IN#	Payment	IN#	Payment

Bank Deposit $ ____________ Total Cash $ ____________

Total Checks $ ____________ Total Cr. Cards $ __________

Total Receipts $ ____________

No. of RO s ________ No. of LOF's ________

FIGURE 9-2 Daily deposits and report for owners.

FIGURE 9-3 Cashier station.

policy should identify positions that will perform each duty and not people. Although such a policy may seem extreme, cases of fraud and theft can be cited when one or more of these conditions has been ignored.

Bank Reconciliation

The information entered into a daily receipt report, such as the one shown in Figure 9-2, should be reconciled with the bank deposits and credit card printouts, which, in turn, should be reconciled with the checkbook and accounting records. These reconciliations must be done monthly and no later than within one week of the receipt of the bank statement by a third person or owner.

When a service facility does not have enough positions or people to perform these duties, then more than one duty may be assigned to a position. In these cases, however, an owner is most often available to perform one or more of the duties. In any case, care must be taken to have checks and balances in place along with regular internal audits of the records. Many trusted employees have not been able to resist the temptation to commit fraud when the checks and balances are not in place and they are under personal financial stress.

Receipts and Transmittal Forms

Facilities should use transmittal forms when money is transferred from one employee to another. When a transfer occurs, the amount of money is entered on the form. The date and amount in cash and checks are noted. Both employees must sign the form and each person retains a copy. A copy of the daily summary of the business deposit should be given to all parties receiving, processing, recording, and depositing money. The employees who handled the money should verify for their own protection that the records are accurate. At unannounced times, the owners and/or managers should conduct internal audits of receipts, transmittals, and deposits to verify they are correct. The audit does not have to check all records but should select days at random for the checks. The spot checks not only protect the resources of the business but also protect honest employees.

Accountants, owners, and managers must always monitor the way the rules and procedures are followed to carry out policies. A slight deviation from the rules and procedures is a tip-off that something is not being done right and the reason may be to violate the company's resources or test the system. Each department manager is responsible for his or her receipts and invoices. In the event of any theft or embezzlement, the manager is always suspect and, ultimately, held accountable when rules and procedures, policy, or a law has been violated.

Credit Cards

Many business transactions involve the use of credit cards. Credit card sales are subject to bank discount fees, meaning that the service facility is charged a fee that reduces the amount of money it receives from the credit card company. The fee is a percentage of the total sale amount, such as 1 to 3 percent of the charge for parts, labor, and sales tax. The cost to the facility also may include a fixed fee to cover other costs (such as the rental of the credit card processing machine and a dedicated phone line for the machine).

A facility must make certain their credit card machine is secure. For example, in one company several employees were running credits for their personal credit cards through the company machine. In other words, their credit card balance was reduced and the company account was charged for it. Therefore, the credit card reports must be checked against the deposit reports, such as the one shown in Figure 9-2, and accounting records.

Bad Checks and Nonpayment

The acceptance of checks carries the risk that will not be accepted by a bank because there is not enough money in the customer's account. A check is returned as NSF (nonsufficient funds) by the bank when the amount of money in a customer's account cannot cover the amount of the check. At the authors' facility, a company was to paid review customer checks (usually only those for large amounts) before they were accepted. If a check bounced after their approval, the company (who approved the check) paid the money owed to the facility. The facility was charged a fee for the service plus another fee for each check that was reviewed. Whether such a service is used by a facility depends on the problems experienced by the facility. Some of the information collected to make this decision would be referred to as nontraditional accounting information.

In some cases, a check may bounce because of customer error, such as writing down a wrong amount in their checkbook. However, if a customer writes a check from a closed account, such as an old bank account no longer in use, this act is not considered an NSF check but theft. In these cases, the customer should be given the benefit of the doubt and notify him/her. If payment is not made immediately after notification, the matter should be turned over to the police, who must determine if the check was written out after the customer closed the account. When a check was written after the customer closed the account, then the service manager or owner may consider pressing charges. For the customer this will likely mean arrest, possible trial, and, if convicted, a fine and incarceration. The authors had to levy such a charge in two situations, both of which resulted in the arrest of the customer and conviction. Restitution was eventually made; however, it took considerable time, during which the company had to cover the cost of the parts and labor incurred in the repair of the customer's car.

Personal Credit

Another problem at service facilities is the extension of unsecured personal credit through a personal note, such as, "I will pay you $X next week." It is not unusual for an inexperienced service facility manager or service consultant to release a car to a customer before they pay the bill. The promise that the customer will be in next week, or later, to pay the bill seems reasonable (as the authors can testify), until they do not show up.

Instead of granting credit when people cannot pay their bill, a facility may use a **mechanics lien** law, which most states have on the books. This law permits the service facility to hold a customer's car until the bill is paid, provided certain conditions are met. The conditions to be met vary from state to state but, in general, the requirements are: (1) the services on the automobile were authorized by the owner of the car (never assume the customer is also the owner); (2) a repair order contained all of the information necessary to be considered a legal contract; and (3) the car is in the possession of the service facility.

When an invoice has not been paid after the automobile has been at a facility for a reasonable amount of time, finance and storage charges usually can be charged for tying up service facility space, capital, and employee time. If the invoice is still not paid and the time that has passed legally indicates that, by default, the property has been abandoned, most state laws allow the service facility to go through a legal process to sell the car for the amount owed. This action usually requires proper documentation and that a process be followed to make the customer aware of the situation, such as the amount of time that has passed since the customer last contacted the service facility and the amount of money owed. Because state laws dictate the exact requirements, an attorney needs to be consulted to get the details.

If credit is awarded and the service facility has to collect money, the customer should be contacted and arrangements for the payment or payments should be made. In most cases, customers will make the payment or payments without issue. However, some customers will be difficult to reach, will not return phone calls, or will make promises that are not kept. When this occurs, a letter, which should be approved by an attorney, must be sent to the customer. This letter should be sent by regular and certified mail. Certified mail has a postcard attached that must be signed by the recipient before he or she can get the letter. Once the postcard is signed, it is sent back to the manager as proof that the customer got the letter. The regular-mail letter is the same letter as the certified-mail letter. While redundant, the regular-mail letter is needed if the customer refuses to sign the post card and the certified letter and post card are returned. Therefore, the only proof the customer got the letter when the manager goes to court (often small-claims court) is that a regular-mail letter was sent at the same time and it was not returned. The judge, once the manager testifies to the process the company uses, will typically assume the customer was notified and will pass judgment to nullify a claim by the customer that he/she was not notified.

An alternative to the above, or as a next step for nonpayment, assuming a customer has still not paid the invoice, is to pay the fees necessary to have the customer's property seized by the sheriff's department and sold at auction to pay the bill. This is a long process that often takes several weeks but eventually results in obtaining the payment or at least some of the money owed to the facility. In the final analysis, however, the best policy is for a facility not to grant credit.

Sales Taxes

Most customer transactions are subject to *state sales tax*. When a state sales tax is collected on a transaction, the business must submit the money to the state at the end of the month or quarter (depending upon state law). Sales tax rates vary from state to state, plus there could be an additional local sales tax charged to customers.

A sales tax is charged for each sale by multiplying the amount of the sale by the sales tax percentage. So if a sale was made for \$100, the tax charged would be \$6.00 (6% × \$100). In some cases, a business may not separate their sales tax from the sale amount when it is recorded, such as the sale of a used part and cash is put into the cash register. Then when the manager counts up the amount of money collected, he/she calculates the amount of sales tax owed. When calculating the amount of sales tax owed on such sales, a mistake is often made in the determining the amount of tax to be paid. For example, assume that a facility had a monthly cash receipt of \$600 and wishes to calculate the amount of sales tax owed. The formula to make this calculation for a 6% sales tax would be as follows:

$$\text{Sales tax owed} = \text{Total Sales} - \frac{\text{Total Sales}}{1.06}$$

$$\text{or } \$600 - \frac{\$600}{1.06} = \$600 - \$566.04 = \$33.96$$

In addition to the sales tax, some customers also may be charged an *excise tax* under special circumstances. For example, some items, such as the refrigerant inventory of large retailers, are federally taxed, and some states have additional taxes on items such as tires. In Pennsylvania, an excise tax on tires is collected to pay for the state cleanup of used-tire dumps.

Some states do not have a sales tax, such as Delaware, Alaska, Oregon, Montana, and New Hampshire; and some customers do not have to pay them because they are *tax-exempt*. Tax-exempt customers are typically nonprofit and government agencies, as well as businesses that resell a service facility's goods or services to their customer. When this happens, the facility must collect sales tax exemption letters from the facility to verify they are tax-exempt. The service manager must keep these documents on file and show them to state revenue officers upon request.

Expense Transactions

An **expense transaction** occurs when a facility pays money for goods and services. Expenses may be divided into different categories, such as expenses that are *directly related to the actual delivery* of a service and sale of goods (technician's wages and parts and supplies that are resold) and indirect expenses, which are incurred *in support of the delivery* of services and sale of goods. The indirect expenses are referred to as **overhead** and include payments such as management and clerical salaries, rent or mortgage payments, gas or oil for heat, water, electricity, telephone, advertising, and so on. The payments made by the facility for parts, wages, and overhead expenses occur at different points in time. For example, employees may be paid at the end of the week, while parts and overhead are paid monthly (typically at the end or beginning of the month).

Some expense transactions earn a discount if paid by a certain date. For example, a parts supplier may offer a service facility a *2/10, net 30* discount on their invoice. This means if the service facility pays the parts supplier invoice by the 10th of the month, they can take 2 percent off the invoice. If they don't pay by the 10th of the month, they have until the 30th of the month to pay the bill in full. Naturally, after the 30th of the month, interest will start to accrue (be charged) on the account. Two percent discounts are a very good deal and can equate to a very high annual return (or reduction in expenses) for the service department in one year. To illustrate, assume a part's invoice from a vendor is $1,000 for the month, and that there is a 2 percent discount (saves $20); therefore the amount owed would be $980.

Internal Control for Spending Money

As noted earlier in the definition and discussion of terms, a budget is an allocation of money that may be spent by a department or unit. These allocations are made for each overhead-expense category, such as salaries, supplies, phone, and so on. The purpose of the allocation is to control the spending of money, because when the amount spent in an expense category equals the amount allocated to the category, no more money can be spent in that category without the permission of the owners or, possibly, upper management. Budget allocations are not usually made for parts and supplies that are resold; however, a comparison between the amount of money received from the sale of parts to the amount spent for parts must be closely monitored. Also technician wages are usually based on the number of full-time and part-time workers needed; however, the amount being spent must be constantly compared to the amount of money being collected from the sale of services.

In the control of expenses, a concern for most automobile service facilities is that they spend hundreds and sometimes thousands of dollars a day on the purchase of parts and supplies. These expenditures are, and should be, primarily credit purchases, while cash purchases should only be made via a petty cash fund (discussed later). Like the internal control over money when it is received, there must be controls when money is spent either through the use of a charge account, credit card, petty cash, or the writing of checks. If not, the loss to a business can drive the company into bankruptcy.

A basic outline for the internal control over expenditures should include at least five stages, which may not be conducted by only one person unless that person is the sole owner of the facility. Of course, these stages may contain more detail depending upon the size of the facility and its organizational structure. An outline of the five stages for internal control is shown in Figure 9-4.

The separation of duties is required for the five stages for internal control. The actions to be taken in each stage are shown in Figure 9-5. For control purposes, one person should not be permitted to conduct the actions in more than one stage with the exception of the owner.

In addition to the above steps, the service manager and owner should audit (check) the process regularly. This means that when walking around the facility, they should notice if purchase receipts are initialed and forwarded, if orders shipped to the facility are being checked, if credit receipts are received and forwarded, and so on. On a busy day in a shop with multiple bays, errors are made or steps are disregarded for the sake of getting a job done. This requires the manager, or owner, to make a more formal and unannounced check of these practices outlined above. In

1) The Purchase of Goods or Services
   a) When money is spent, someone must initiate an expense transaction for the facility by placing an order by phone or computer, or making a direct purchase at another business.
2) Receipt of Goods
   a) When the order is received at the facility or an item is picked up at a store, a receiving packing receipt (a piece of paper or cardboard that presents what has been delivered) or store receipt (showing what was purchased) must be obtained.
3) Receipt of Invoice
   a) Invoices are received for payment.
4) Payment
   a) A check is written to pay for the purchases.
5) Bank Reconciliation
   a) The bank statement is received and all deposits (including credit card deposits), canceled checks, and bank balance must be checked against the checkbook, credit card reports, and accounting records.

FIGURE 9-4 Five stages for internal expenditure control.

1) The Purchase
   a) Only approved employees should be permitted to make credit or petty cash purchases. Preferably, the vendor of parts stores will only permit employees on an approval list to make the purchases.
2) Receipt of Goods
   a) When a purchase is received, either in person at a store or by delivery to the facility, the person making the purchase or receiving the order must verify that the item (or items) received is correct. The delivery/packing slip or store receipt must be initialed by the employee who receives the goods to indicate the order was received and was accurate. The receipt must then be forwarded to the accounts payable clerk, bookkeeper, or owner.
   b) When a part, core, or used part is returned and credit is received, a receipt for the part must be received (or made up by the service manager) and then forwarded to the person who checks invoices.
3) Receipt of Invoice
   a) When invoices are received and before payment can be made, a copy of the receipt of the purchase must be attached to the customer invoice by the accounts payable clerk, bookkeeper, or owner for comparison.
   b) Credits for the return of parts and cores must be compared to the charges/credits on the invoice.
   c) After the invoice is checked an amount is approved for payment (for example, a charge may not be correct and after calling the vendor, the invoice balance may be adjusted). The invoice is then forwarded for payment.
4) Payment
   a) The invoice showing that the charges that have been verified for payment and the credits received is reviewed by the person who writes the checks. A check is written and signed by the business manager or owner.
5) Bank Reconciliation
   a) The treasurer or an owner of the facility must reconcile the bank statement. The deposits (including credit card deposits), canceled checks, and balance shown on the bank statement must be checked against the checkbook, credit card reports, and accounting records. All bank charges and NSF checks must be subtracted from the balance in the checkbook and entered into the accounting records.

FIGURE 9-5 Actions taken in the expenditure stages.

some cases, checks on performances are the only evidence an employee has that the practice is important.

The authors can testify from personal and painful experience that there are more ways a facility can lose money than imagined when expenditures are not controlled. When duties are not divided, when receiving slips and receipts are not verified, when invoices are not confirmed, and when bank statements are not reconciled, the business is at risk. For example, in one case the service manager, who was authorized to make purchases, was buying parts for a small repair business he had on the side. The parts were charged to the owner's business. The verification of the charges on the invoices immediately revealed the theft. In another case, a parts store was charging the facility for parts that were never purchased. They claimed they were simple errors; however, another parts store was used for future purchases. A company cannot do business with another company that does not maintain controls over their operations.

Petty Cash

A **petty cash account** is when money is made available for small purchases. For example, a mail delivery may have $.30 postage due, $2 may be needed to send a registered letter, or $15 may be needed to purchase a cartridge for the printer. The owners or manager of the facility must set a limit on the amount of money that may be spent on a purchase from the petty cash fund, such as a maximum of $20 or $25. When a purchase is made from the fund, the receipt is placed in the lockbox containing the petty cash money. When the fund gets below a set amount, say $10, it is replenished by giving the receipts to the person who writes the checks. This person then makes out a check to *Cash* for the amount of the receipts, cashes the check, and gives the money to the person in charge of the petty cash.

Inventory Control

Major expenditures also are made by automobile service facilities for the advance purchase of goods and supplies to be resold, such as tires, oil, antifreeze, oil filters, air filters, and so on. When these materials are received, they become classified as inventory. Maintaining inventory records is also important to the financial security of a facility.

When items or materials are purchased, inventory records must be created; for example, if 100 tires are purchased, the date of purchase, make and model, size, type, cost, and number of tires received must be recorded. As tires are sold, the inventory sheet must be updated. Then the service managers should check or have the inventories taken on a regular basis. In the case of bulk purchases, such as oil, the records of the amount delivered, number of changes made, and the estimated amount left in inventory should be checked and recorded. As the authors learned the hard way, inventories cannot be ignored. In fact, when a service consultant or manager who has access to inventories and their records gives notice that he/she is leaving, an inventory must be (not should be) taken!

Final Thoughts

Having control of the financial operations in a service facility is the first step in financial management. If managers do not have the income and expense operations properly set up, financial plans cannot be relevant because of control issues. For example, a plan to carry water in a bucket is useless if the bucket has holes in it.

Assuming the internal controls and proper practices are in place, the following chapters describe the contents of financial statements, labor, and mark-up rates, and how this information is used to make management decisions and carry out operational plans. When a sound plan is put into operation, proper controls must be in place or the plan, regardless of how good it is, will not succeed. As an illustration, the following case of a successful automobile dealer is presented.

A gentleman owned a very successful dealership for many years in a small city. The owner, however, did not think it was important for checks and balances to be in place because he had operated his business for years without any problems. He made a lot of money (so he thought) and had loyal (so he thought), longtime employees. Then one day, a car manufacturer's district service representative (personal friend of the authors) visited his store to tell him that his order for new cars for the upcoming year could not be filled. The representative explained that his account had been audited and he no longer had any credit. This obviously shocked the owner.

The representative felt quite bad about the situation and assisted him in a review of his financial records. Then he had to tell him he found that the trusted bookkeeper of 20-plus years had failed to keep his payments up-to-date and he did not have a balance in his checking account. Obviously, the bookkeeper did not pay the manufacturer and embezzled the dealer's money. The dealer did not have the money that he thought he had. In fact, he did not have any money and a lot of debt. This, of course, caused the dealership to close. The owner, who had a top-flight reputation in this small city, was so embarrassed and afraid of the damage that would be done to his image and his family, he did not press charges. We are not sure about the eventual fate of this owner; however, it is not as rare a case as one might think.

Review Questions

1. What is a transaction?
2. What is the purpose of accounting?
3. What is the difference between income and revenue?
4. What is the difference between expenses and expenditures?
5. What is the purpose of inventory control?
6. What do proper internal-control procedures for receipts and expenditures look like?
7. Explain the difference between gross profit and net profit.
8. What is a cash transmittal form and why is it used?
9. Give an example of nontraditional accounting information and explain how it could be used.
10. What does 2/10, net 30 mean?

CHAPTER 10

ACCOUNTING PROCESS AND FINANCIAL STATEMENTS

LEARNING OBJECTIVES

Upon reading this chapter, students should be able to:

- *Explain the difference between real and nominal accounts.*
- *Define assets, liabilities, and equity accounts.*
- *Describe the three sections, accounts and its purpose of the income statement.*
- *Describe the three parts, accounts, and purpose of the balance sheet.*
- *Calculate straight line depreciation.*
- *Explain the relationship between the cost of sales to sales accounts and the account title given to their difference.*
- *Define book value.*
- *Explain the difference between cash basis and accrual basis accounting.*

Introduction

As explained in the last chapter, business transactions are entered into the accounting system. The accounting system then processes the transactions to produce financial statements. The flow of information about the transaction into the accounting process is illustrated in Figure 10-1. The purpose of this chapter is to discuss the financial statements that service facility owners and managers use to monitor company operations in reference to its objectives as set forth in its strategic business plan.

A strategic business plan with objectives is not difficult to prepare. The key is knowing what to monitor and how to oversee the process as the employees work toward the objectives. However, managers often believe that monitoring business activities is time-consuming and complicated. These negative feelings often arise because a review of financial information requires a manager to plan and make tough decisions. The alternative, of course, is to *fly blind.* Without accurate reviews of solid, meaningful information, managers are limited to becoming caretakers of problems that come through the door. By default, the manager no longer leads the business but the business leads the manager. This means when the owners discover the business is not going in the right direction and actions are needed to reduce company losses, the manager becomes expendable.

To make decisions, accurate financial data is needed. To obtain accurate financial data, the handling of and amount of money received and spent must be monitored, otherwise the financial statements will not be accurate. When the numbers are off and managers are not aware of it (as happens when they do not monitor the process), they are more likely make wrong decisions or reach improper conclusions, which are called *material errors*. To illustrate this point, a student of the author's reported

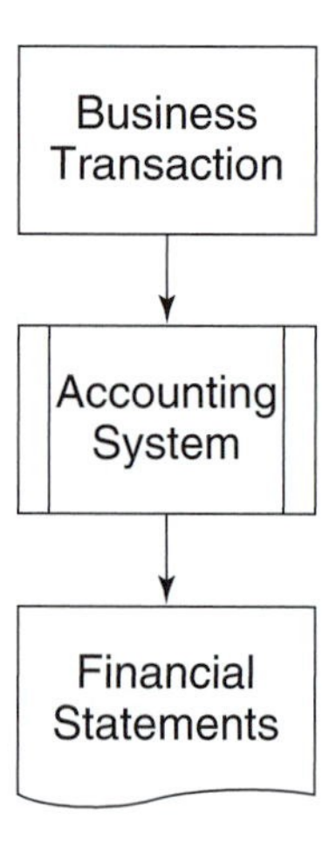

FIGURE 10-1 Transactions to financial statements.

that he worked Saturdays for a service facility that did not have a manager or owner on duty; the shop foreman was left in charge. The shop foreman collected all of the Saturday cash sales and put it in his pocket. At the end of the day, the shop foreman and the technicians would split up the cash. When the owner found that Saturday work was not profitable, he asked the manager why the shop was open. The manager did not know the answer and asked the shop foreman about cutting costs by not having so many workers on Saturday. The foreman told the manager that Saturday was the day they caught up on work and that all of the workers were needed. The manager did not know that more income came in on Saturday than was reported. Failure to have controls that allowed the process to be monitored by the manager meant that Saturday work was potentially one of the most profitable days. The result was that the financial statements were not accurate because they showed a slight loss every Saturday. However, management did not feel the slight loss was material and continued the Saturday work schedule. Of course, the manager was wrong and should have caught the problem by looking at collaborative data, such as expenditures for parts, supply inventory, and use of oil (discussed in the last chapter), as well as some of the comparisons to be discussed later in this book.

The Accounting System

When transactions are entered into the accounting system, they are placed in different account classifications. First, the accounts are divided into two types: *real accounts* and *nominal accounts*. The balances in real accounts are carried over from year to year; for example, the cash balance at the end of the fiscal year (the fiscal year is when the business closes the books and pays its end of the year taxes) will become the balance at the beginning of the next fiscal year. Nominal accounts, however, are temporary accounts because they are used for one fiscal year. For example, a sales account records sales made by the service facility (called sales receipts) for the year, which may total a few hundred thousand dollars. The balance in the sales account will be zero once the new fiscal year starts. Likewise, an expense account such as rent will record charges for a fiscal year. At the end of the fiscal year, this account is closed and the new fiscal year starts with a zero balance when it opens again for new entries.

Real Accounts

The three classifications for real accounts are:

1. Assets
2. Liabilities
3. Owner's or **stockholder's equity**

Assets are property or rights that have value. Assets accounts are used to record:

- Cash and investments
- What is owed by others (called accounts receivables) and the business has the right to collect
- What the company owns and uses to conduct current and future business (such as buildings equipment, supplies, and inventory)

Assets can further be divided into two subgroups, which are: **current** and **long-term assets.** Current assets are cash or anything that can be converted to cash in a short period of time without disrupting business operations (for example, selling inventory at cost or a loss to recover the money). Long-term assets are property and equipment (often called fixed assets because they have more than a year of useful life) that cannot be converted to cash very easily or quickly but have a future economic benefit.

Liability accounts record what a business owes. Liabilities may be divided into two subgroups, which are **current** and **long-term liabilities.** Current liabilities represent short-term credit arrangements to be paid within the fiscal year. For example, an amount of money borrowed from a line of credit the bank gives the service facility when they do not have enough cash to make their bills is a current liability. Long-term liabilities are credit arrangements that will take longer than a year to pay off, for example, a loan to buy expensive equipment or a mortgage to buy a building.

The owner's or stockholder's equity account is the amount of money the owners or stockholders would have after the assets would be used to pay off all of the liabilities. A company is considered liquid when the assets are worth more than the liabilities. To encourage investment back into the service facility so that it can stay up-to-date (for example, to buy new equipment) and grow (for example, to hire more employees), the owner's equity section has a special account called the retained earnings account (also known as earnings retained by the business). This account has money placed in it when the company has a profit at the end of the fiscal year. As explained in Chapter 2, however, proprietorships, partnerships, and S-corporations are considered pass-through entities that do not have retained earnings accounts (although the owners may choose to buy investments that will be an asset owned by the business). Rather, pass-through entities require that the profit be passed to the owners, who pay tax on it, and then at their discretion choose to reinvest the money back into the business.

Nominal Accounts

The two types of nominal accounts, which were defined in the last chapter, are *income accounts* and *expenses accounts.* These accounts should be thought of in terms of their effect on owner's (or stockholder's) equity. Specifically, expenses have a negative affect on equity while income adds to it. Therefore, at the end of the fiscal year, when the amount of money received (income) exceeds the amount spent (expenses), there is a profit and the balance in the owner's or stockholder's equity will increase. For example, if expense = $60,000 and income = $80,000, the effect on the owner's (or stockholder's) equity would be an increase of $20,000 (the profit earned for the year).

The Double Entry Formula

Accounting uses a system to keep the books balanced. The system is called double entry accounting, and the equation is as follows:

Assets = Liabilities + Owner's Equity.

This formula represents the way the accounts appear on a balance sheet, but the equation is more easily understood as follows:

Assets − Liabilities = Owner's Equity.

In this case, assets (what is owned) minus the liabilities (what is owed) equals what the owners (or stockholders) would have left over after all debts are paid. For example, if the assets = $97,000 and the liabilities = $20,000, the owner's or stockholder's equity would equal $77,000. To keep the equation balanced for each transaction at the service facility, double entry accounting rules dictate that a change (or entry) must be made in at least two accounts.

Capital Assets and Depreciation

Capital assets are fixed assets such as buildings, land, and equipment. To spread the cost of a capital asset over its expected useful life (except land), it is **depreciated** annually. For example, one method of depreciation (called *straight line depreciation*) is to take the cost of the asset and subtract its estimated salvage value at the end of its useful life. The difference is then divided by the number of years of its estimated useful life. For example, if a machine cost $21,000, had an estimated salvage value of $1,000, and was estimated to have a life of 10 years, then the annual depreciation would be $2,000 or ($21,000 − $1,000)/10 years = $2,000. The amount of the depreciation does not represent an amount of money spent each year for the asset or money set aside in a savings account. It is simply a means to generate an expense that will recover the amount of money paid for an asset to compensate the business for wear on the asset. Therefore, in this example, the annual depreciation expense would decrease the business profit by $2,000 to compensate the business for the use of equipment. Because the owner is taxed on the business profits, depreciation expense would reduce the owner's taxable income by

$2,000. However, this is a simplified example, and in reality for tax purposes a depreciation method called the Modified Cost Recovery System is used. There are many rules and a tax expert must be consulted to ensure the proper procedures are followed.

Financial Statements

There are many different types of accounting documents. The two financial statements covered in this text are the income statement for nominal accounts and the balance sheet for real accounts. Figure 10-2 shows the relationship of these two statements and their accounts relative to the daily transactions at a service facility. Introduced in the previous chapter, the income statement presents the income and expenses (nominal accounts) for a period of time, such as the month of March or the first quarter of the fiscal year. Modern accounting software can help produce an income statement for any day or period of time. In fact, if an owner or manager wishes, the account balances in an income statement can be checked at the end of every business day provided the proper accounting entries have been made. Obviously, with the ability to track the effects of incomes and expenses instantly, this statement is most relevant to a manager's and owner's oversight of daily operations.

For a manager, the main purpose of the income statement is to provide information about how much money was received, how much was spent, and whether the profit meets the objectives in the strategic busi-

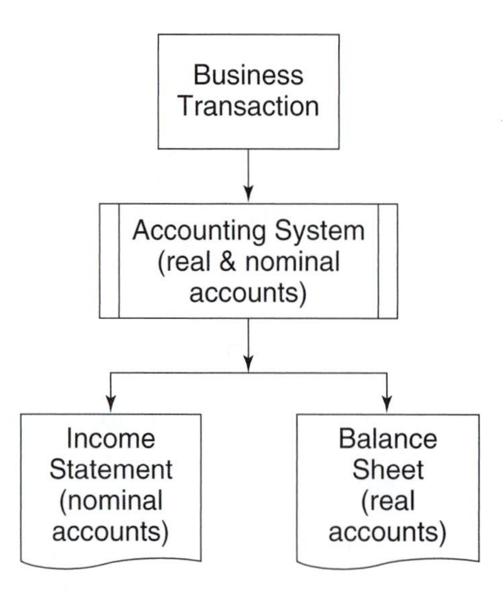

FIGURE 10-2 Transactions to income statements and balance sheet.

ness plan. The income statement covers a period of time from the first to the last day in a period. To use the income statement as a benchmark, income, expense, and profit targets must be set to determine whether the business will meet its financial objectives.

For a service manager, the balance sheet presents the value of all company assets and liabilities plus the owner's equity accounts (the real accounts). The amounts given in the balance sheet are as of a specific date (such as the last day of the period) and not a period of time. Of course, the balances in these accounts will change every day and with the use of accounting software, a balance sheet could be created every day. Because the balance sheet presents the real accounts for a business, it is of great interest to owners, stockholders, investors, suppliers, and banks, among others. When a loan is requested, a bank, for example, wants to know what the business is worth (net worth equals owner's equity).

The Basic Income Statement

An income statement can be fairly complex because there are many different accounts, which are called line items, for income and expense (nominal) accounts. However, regardless of how many line items are on an income statement, there are three parts: the income section, the expense section, and net income. A basic format for an income statement is shown for Bob's Auto Repair in Figure 10-3. This income statement presents the period of time covered (the month of February) and the three

Bob's Auto Repair
MONTHLY INCOME STATEMENT
FOR PERIOD ENDING ENDING FEBRUARY 28, 2005

INCOME			
	Labor to fix cars	4,000	
	Storage Fees	300	
	Total Revenue		4,300
EXPENSES			
	Salaries	2,000	
	Rent	1,000	
	Gas, water, electric	500	
	Advertising	300	
	Total Expenses		3,800
NET INCOME			500

FIGURE 10-3 Basic income statement format.

sections. Therefore, because it shows the bills paid and income earned in February, it is a period statement.

Full Income Statement

An income statement typically will show a number of income and expense line items that a business uses to keep track of the way its money was earned and spent during the year. The number of line items to be used depends on the business and the preferences of the owners and managers. In other words, there is no list that must be used. The intent is to be able to produce a report that will be acceptable to the owners or stockholders and tax collectors (the Internal Revenue Service, or IRS).

An income statement, therefore, presents the amounts received and paid. It also may present the amounts earned or spent (cumulative amounts) since the beginning of the fiscal year and the percentages of each expense compared to the income earned. For most businesses, a fiscal year runs from January 1 to December 31; however, a company may use a different 12-month period with the permission of the IRS. Figure 10-4 shows an income statement with the percentages and cumulative totals amounts since the beginning of the fiscal year, which was January 1.

The income statement in Figure 10-4 for Renrag Inc. shows the cost of sales, which is the amount spent to service customers' automobiles (both parts and technician's labor), for March to be $10,636, with $27,067.97 for the year. This amounted to 38.1 percent of the total income earned for March and 37.6 percent of the income earned for the year. The gross profit (total income less cost of sales) was $17,311.03 (61.9 percent) for March and $44,925.22 (62.4 percent) for the three months. The question for managers and the owners is whether this is good enough. The answer depends on the bottom line, which is the profit for the month (a loss of $917.88 or –3.28 percent) and the period (a profit of $6,295.37 or 8.75 percent) as well as the strategic business plan and the benchmarks set for the end of the first three-month period. If the business plan expected a 12 percent net income at the end of the first three months, then management must review the sales, cost of goods, and each of the expenses for each month in the period and especially the month of March.

Consequently, the income statement can present a considerable amount of information when it is set up to record the sales and expenses relevant to the operation of the facility. Further discussion of the analysis of the statement is included in later chapters in this book. For example, further examination of the percentages reveals that two expense line items in March were higher than the year to date. These line items were interest expense and management salaries and wages. Research is needed to determine why these items are greater for March than the average year to date. For instance, what were they for the last fiscal year?

RENRAG INC.
INCOME STATEMENT
Period Ending
3/31/06

Note: When using the Excel spreadsheet provided on the Delmar Website, some line items use an equation to calculate the number seen by the user. For ease of use, all numbers are automatically rounded to one or two decimal places by the program. If the user were to use a calculator to add the line items in a column together, the total of the rounded numbers may not appear accurate when compared to the program's total of the column. The reason is because Excel does not use rounded numbers when adding. Rather, the program adds the calculated numbers before rounding, then rounds the number after adding them together. For example, a total of 37.6 is shown in the column and if a caculator is used to total the numbers, the total of the rounded numbers in the column is 37.7. In reality the 37.6 number is more accurate because it was added together by the program before rounding.

	MARCH	Cummulative	Year to Date	Cummulative
INCOME				
Labor Sales	12,140.25	43.4%	30,596.37	2.5%
Sales Oil	539.00	1.9%	1,244.96	1.7%
Sales-Parts and Accessories	14,952.88	53.5%	38,317.14	53.2%
Sales-Tires	421.72	1.5%	1,356.12	1.9%
Sales-Other	0.00	0.0%	506.95	0.7%
Other Income	30.79	0.1%	103.42	0.1%
Interest Revenue	0.00	0.0%	0.00	0.0%
Sales Returns and Allowances	<136.77>	-0.5%	<136.77>	-0.2%
Sales Discounts	0.00	0.0%	0.00	0.0%
Total Income	27,947.87	100.0%	71,988.19	100.0%
COST OF SALES				
Technician Wages	4,076.90	14.6%	8,497.75	11.8%
Overtime	39.38	0.1%	39.38	0.1%
Bonus - Tech	551.35	2.0%	710.65	1.0%
Cost of Goods Sold - Oil	479.00	1.7%	1,119.00	1.6%
Cost of Parts	4,637.30	16.6%	14,753.75	20.5%
Cost of Goods Sold - Tires	294.46	1.1%	924.85	1.3%
Cost of Sales - Other	558.45	2.0%	1,017.59	1.4%
Purchase Discounts	0.00	0.0%	0.00	0.0%
Total Cost of Sales	10,636.84	38.1%	27,062.97	37.7%
Gross Profit	17,311.03	61.9%	44,925.22	62.4%
EXPENSES				
Advertising Expense	695.20	2.5%	2,200.65	3.1%
Auto Expenses	247.13	0.9%	282.28	0.4%
Bad Debt Expense	0.00	0.0%	0.00	0.0%
Bank Charges	186.69	0.7%	501.47	0.7%
Bonus - Manager	0.00	0.0%	0.00	0.0%
Cash Over and Short	0.00	0.0%	0.00	0.0%
Charitable Contributions Exp	0.00	0.0%	0.00	0.0%
Consultants	50.00	0.2%	50.00	0.1%
Management Salaries and Wages	3,596.25	12.9%	6,996.25	9.7%

(continued)

FIGURE 10-4 Full income statement with percentages.

	MARCH	Cummulative	Year to Date	Cummulative
Managers Commission	908.28	3.2%	1,635.52	2.3%
Depreciation Expense	0.00	0.0%	0.00	0.0%
Dues and Subscriptions Exp	0.00	0.0%	243.20	0.3%
Electric	290.22	1.0%	865.13	1.2%
Employee Benefit Programs Expense	460.20	1.6%	1,677.10	2.3%
Gifts Expense	0.00	0.0%	0.00	0.0%
Heat-Gas	161.78	0.6%	514.48	0.7%
Heat-Oil	0.00	0.0%	150.00	0.2%
Insurance Expense	0.00	0.0%	1,140.45	1.6%
Interest Expense	2,877.09	10.3%	3,819.71	5.3%
Laundry and Cleaning Expense	148.40	0.5%	482.30	0.7%
Legal and Professional Expense	975.00	3.5%	1,075.00	1.5%
Licenses Expense	0.00	0.0%	0.00	0.0%
Loss on NSF Checks	0.00	0.0%	0.00	0.0%
Meals and Entertainment Exp	0.00	0.0%	0.00	0.0%
Office Expense	29.32	0.1%	145.86	0.2%
Payroll Tax Expense	938.36	3.4%	1,829.17	2.5%
Penalties and Fines Expense	0.00	0.0%	0.00	0.0%
Pension/Profit-Sharing Plan Ex	0.00	0.0%	0.00	0.0%
Postage Expense	38.33	0.1%	163.97	0.2%
Rent Expense	2,100.00	7.5%	6,300.00	8.8%
Equipment Lease Expense	699.87	2.5%	2,099.61	2.9%
Freight (FOB) charges to ship parts	0.00	0.0%	5.00	0.0%
Repairs & Maint. Expense	457.35	1.6%	874.60	1.2%
Repair Manuals	0.00	0.0%	0.00	0.0%
Security	0.00	0.0%	92.22	0.1%
Shipping Expense	0.00	0.0%	0.00	0.0%
Supplies Expense - Shop	77.51	0.3%	275.05	0.4%
Tax Expense	0.00	0.0%	0.00	0.0%
Real Estate Taxes	570.00	2.0%	570.00	0.8%
PA Capital Stock Tax	0.00	0.0%	0.00	0.0%
Business Privilege Tax	617.14	2.2%	617.14	0.9%
Telephone Expense	586.02	2.1%	928.34	1.3%
Tools Expense	89.26	0.3%	768.99	1.1%
Training	0.00	0.0%	137.00	0.2%
Travel Expense	605.88	2.2%	684.98	1.0%
TV Cable	30.99	0.1%	92.97	0.1%
Uniforms	145.12	0.5%	469.34	0.7%
Waste Removal	60.00	0.2%	250.00	0.3%
Water & Sewage	18.54	0.1%	102.12	0.1%
Other Expense	568.98	2.0%	589.95	0.8%
Gain/Loss on Sale of Assets	0.00	0.0%	0.00	0.0%
Total Expenses	18,228.91	65.2%	38,629.85	53.7%
Net Income	<917.88>	<3.28>	6,295.37	8.75

FIGURE 10-4 *continued*

Merchandise Inventory Worksheet

In the last chapter, the importance of inventory control was discussed. When the inventory is maintained as suggested, the cost of the inventory sold must be included in the cost of sales section. In Figure 10-4, three accounts are used to record the cost of the inventory sold: Cost of Goods Sold—Oil; Cost of Goods Sold—Tires; and Cost of Goods Sold—Other. The last account would be used to charge customers for a variety of items on inventory, such as bolts, tubing, pipes, and so on.

Merchandise Inventory Worksheets should be maintained for each Cost of Goods Sold account, such as oil and tires. When the inventory consists of goods that are specific and can be tracked, such as tires, an inventory worksheet must record each item and its cost when it is purchased and remove it when it is sold. For instance, a tire in an inventory will have a size, type, and cost. When it is sold, the Cost of Goods Sold—Tires will record the amount that was paid for the tire.

The calculation of the cost of goods that were bulk purchases, such as oil and antifreeze, requires a different worksheet. These worksheets record the bulk purchase of the inventory (beginning inventory) and the amount or estimated amount remaining at the end of the period (ending inventory), which is usually the end of the month. The difference between the beginning and ending inventory gives the amount sold. Figure 10-5 presents an oil inventory work sheet.

When oil is purchased and placed in the tank, it is added to the beginning inventory. At the end of the month, the inventory remaining in the tank is the ending inventory. The beginning inventory for the next period (or month) is the ending inventory of the previous month. For example, Figure 10-5 presents an oil inventory worksheet for Renrag for the month of March. In the worksheet, the cost of the purchase and the number of gallons bought are entered into the worksheet. At Renrag, the cost of one gallon of motor oil was $4 ($800/200 gallons). At the end of the month, there were 75.75 gallons remaining, which is valued at $303 (75.75 × $4). The cost of the oil sold was, therefore, $497, which is the amount of the Cost of Goods Sold—Oil in Figure 10-5.

The same method is used when other supplies are purchased. Although there are several methods that can be used to add to the inven-

Oil Sales	Dollars	Gallons
Beginning Inventory	$800.00	200
Purchases	$0.00	0
Merchandise Available for Sale	$800.00	200
Ending Inventory	$303.00	75.75
Cost of Goods Sold	$497.00	124.25

FIGURE 10-5 Merchandise inventory worksheet for oil.

tory and to calculate the cost, the easiest method is to simply average the cost of the amount of the merchandise available for sale. Using the oil example above, assume that in April another 100 gallons of oil are purchased at $4.25 per gallon. The dollars spent (cost) for a gallon of oil would then be calculated as follows where the average cost of a gallon of oil was $4.14:

Beginning Inventory	$303.00	75.75 gallons
Purchases	$425.00	100.00 gallons
Merchandise Available for Sale	$728.00	175.75 gallons
Cost of Goods-Oil	$728.00 / 175.75 = $4.14 per gallon	

Income Statement Expenses

As shown in the Renrag income statement (Figure 10-4), the list of accounts used by the facility was extensive. A definition for each item is given below to exhibit what each was used to record; however, understand that the Internal Revenue Service (IRS) has the final word on whether an expense can be used to reduce net income because the company or owners must pay taxes on the net income. If expenses are high, net income will be lowered and less taxes will be paid. For example, a net income of $1,000 will result in $300 paid in taxes if the owner is in the 30 percent tax bracket (plus employees' share of Social Security and Medicare). When expenses are deliberately overestimated or false receipts are used to enter expenses into a line item account, it is called tax evasion. Such illegal activities can result in fines, penalties, interest paid on errors, and possibly incarceration. When expenses are permitted by the IRS code and an owner (usually with the help of a tax accountant) uses them to reduce taxes, it is referred to as tax avoidance and is legal. Therefore, as the following list is examined, recognize that a business must have a receipt for every expense transaction to prove it was legitimate and the IRS code must be consulted to determine what may be used in the expense account as an income deduction.

Advertising:

Cost for promoting the business in newspaper, radio, television advertisements; through the preparation and distribution of flyers and handouts; billboard notices; and the use of other mediums such as Web pages to attract customers. Advertisements through charities (such as a service for disabled people), nonprofit organizations (such as a church), or public groups (such as the high school football program) may be classified as a charitable donation and not advertising.

Auto Expense:

The cost to operate a personal or business-owned vehicle for business purposes. The federal government allows a business to charge a straight fee (such as 37¢) for every mile driven; however, a log must be maintained for reimbursements.

Bad Debt Expense:

The loss from sales never collected.

Bank Charges:

Fees and charges varying from the purchase of checks to checking account fees.

Bonus:

An amount paid to an employee that is greater than his/her earned salary or wage. For example, a manager may be paid a fixed amount of money or a percentage (variable amount) when the monthly net income exceeds a predetermined amount or a $100 bonus for every $500 of net income earned over the first $2,000 of net income.

Cash Over or Short:

When cash collected does not equal the sales or cash register receipts.

Charitable Contributions:

Money given to legitimate charitable and nonprofit organizations. The owner or committee must examine each request to ensure the organization and payment qualifies as a legally recognized donation according to the IRS code.

Consultants:

Payments to people who are not employees and hired by the service facility for assistance on technical and business operations and problems. The consultant is recognized as a business, such as a proprietorship, and as a result the facility does not withhold income taxes or make Social Security payments. At the end of the year, if the consultant is paid more than an amount specified by IRS, a form 1099 must be sent to them.

Management Salaries and Wages:

Salaries are payments for a fixed amount money paid at the end of a period of time, such as a week, two weeks, or a month. Wage payments are made for time worked, such as a set dollar amount paid for each hour worked. Because managers do not provide direct services to customer automobiles, their salaries and wages are entered as an expense item and not as a Cost of Sale. This difference affects the amount a facility pays for workman's compensation. In larger facilities, there might be different expense accounts for managers, service consultants, clerical workers, and so on. Also, the amount paid to managers often depends on the size of the business and its gross sales because complex organizations are more difficult to manage.

Commissions:

Commission payments are typically a percentage of sales, such as a 6 percent commission on the sale of goods. A commission may be paid as opposed to a salary; for instance, a manager, who will receive 7 percent of the gross sales, or a manager who may earn a commission in addition to a salary, such as $400 per week plus 3 percent of the gross sales per month.

Depreciation:

The recognition of the cost to use of a capital asset, such as equipment and buildings, over a fiscal year.

Dues and Subscriptions:

The payment for magazines for the waiting room, technical bulletins and journal, and software updates needed by the service consultant and technicians. In addition, payments may be for memberships in business and professional organizations, such as the Chamber of Commerce or the Automobile Service Association (ASA).

Electricity:

Charges for the use of electric power for lights, air conditioning, equipment, and possibly heat.

Employee Benefit Programs Expense:

Employer's cost to provide benefits to employees, including medical insurance and dental and eye care. Although benefits are very expensive, they can be beneficial in the recruitment and retention of top employees. In addition, medical insurance can prevent the abuse of the employer's workman's compensation insurance. For example, an employee was injured on the weekend but waited until he came to work in order to fall over and file a workman's compensation claim. With paid medical insurance, the worker could have gone to the emergency room over the weekend to obtain treatment.

Gift Expense:

The cost of gifts to employees, customers, and suppliers on holidays.

Heat/gas or Heat/oil:

The cost to heat a facility or portions of it. When a shop uses both types of fuel, two line items may be used. In the author's experience, an excellent source of heat is a waste-oil heater, which burns used motor oil and transmission fluid.

Insurance Expense:

Insurance is needed:

- To protect the property and building in case it is damaged by fire (insurance also may be purchased for protection from other events, such as floods and storm) and to protect the business in the case someone is injured on the property.
- By law for workman's compensation (if an employee gets hurt on the job).
- To insure customer car when in the service facility's possession.
- To make sure that if the work is not done properly, someone is injured or property damaged, the customer's claim will be paid (*liability insurance*).

Without proper insurance coverage, the owners put the company assets as well as their entire investment at risk.

Interest Expense:

Interest on loans taken to buy equipment, mortgage or equity loans to add space to the building, or finance cash shortfalls. When a loan payment is made, the portion of the payment for interest must be recorded as an expense (it reduces the taxable income) because the amount for the payment of the principal goes to reduce the amount of the loan, which is a liability (real account) and does not reduce taxable income.

Laundry Expense and Cleaning:

Payments for cleaning employee uniforms and the service facility (bathrooms, waiting area, etc.). It is common for a service facility to contract one or both of these services out instead of hiring employees to do them.

Legal and Professional Expenses:

Costs for attorneys to review or prepare contracts or legal representation in the event of a problem. Other professional expenses include tax accountants and financial specialists.

Licenses:

Payments made to a state or local government for the conduct of business, such as a fee to obtain a state inspection license from the state and for vehicle licenses. In some cases, the service facility may assist their technicians when they need to renew a special license to conduct some jobs, such as air-conditioning repairs.

Loss on NSF (nonsufficient funds) Checks:

When an NSF check is returned (check bounces), the amount of the check (sale) is entered into the bad debts Expense account and the fee charged by the bank is entered as a separate expense (loss on NSF Checks). The separate entry keeps a record of the fee for the bad check in case it can be recovered. In addition, the charge for an NSF check should not appear as a Bank Charge because it is for a bank service and will, therefore, inflate the amount of the bank charges. This may be critical when considering a change in bank services.

Meals and Entertainment:

An expense often abused and carefully reviewed by auditors and the IRS. Only certain types and portions (see the IRS directives) of meals and entertainment expenses can be written off if they are directly related to business activities.

Office Expense:

Office expenses are extensive and may include pens, paper, paper cups, coffee and coffee filters, invoices, printer cartridges, cleaning supplies, and many other items.

Payroll Tax Expense:

The government requires employers to make payments into Social Security (currently 6.2 percent of wages, commissions, bonuses, and salaries paid up to a maximum amount that increases every year) and Medicare taxes (1.45 percent on all wages, commissions, bonuses, and salaries paid with no maximum) for each employee. The employee also must pay 6.2 percent into Social Security and 1.45 percent into Medicare. The employer must withhold the Social Security and Medicare tax from the employee's pay and with the employer's share, send it into the federal government. The line item for the Social Security tax may appear in some statements as FICA, which stands for the Federal Insurance Contributions Act. In addition, the federal, state, or even local government may have taxes the employer must pay such as the Federal Unemployment Tax of .08 percent on the first $7,000 paid to each employee.

Penalties and Fines:

Although this line item may appear on an income statement, not all penalties and fines are deductible as an expense. A tax specialist and the IRS code should be consulted before taking either as an expense.

Pension/Profit-Sharing Plan:

To recruit employees (technicians, managers, service consultants, and parts specialists, among others) a benefit package may include a retirement or profit sharing plan. Although not all businesses offer these to employees, especially smaller operations, they are a popular expense. Retirement plans typically are voluntary and are referred to as a 401K plan, in which employee may deposit a portion of his/her earned income into an account before taxes along with a matching portion from the employer. Not all employees participate; therefore, it is often considered a voluntary plan. A profit-sharing plan simply shares a portion of the profits with employees, who must pay taxes on the earnings that are an expense to the business.

Postage:

This line item is for stamps, express mail services (FedEx, USPS Express service), and other mail-related services.

Rent Expense:

This account is for the rent of any space needed to conduct business. A rental fee may be charged when the owner of the facility also owns the building in another business. There may be several reasons an owner of a business and building may have separate companies; for example, the owner may wish to sell one of them. Rent also may be paid for equipment and special tools the business needs for a particular job or a limited amount of time.

Equipment Lease Expense:

A common method for established service facilities to obtain expensive equipment is to lease it. (A new service facility with fewer than two years of credit history often cannot obtain a lease.) Lease payments are usually affordable amounts to be paid each month for the equipment. The entire lease payment is a deductible expense and then a small amount included in the last lease payment is regarded as the purchase payment. Therefore, the service facility can charge all of the lease payments off as an expense; however, the cost of the equipment cannot be depreciated except for the amount in the last payment, which might be $1. In addition, if the business is sold, the leasing company may take the equipment back.

Freight Charges to Ship Parts:

A facility may ship parts to be repaired (such as a digital dashboard and radios), or sell parts and ship them to customers. Although some consider shipping to the customer (see Shipping Expenses below for items shipped to the service facility) a cost of sales, it is really an expense. When the service facility desires to recover the cost to ship the item (sometimes a handling fee is also charged), the cost (typically with minimal markup, if any) will be recovered under Income—other (the amount is invoiced to the customer). Therefore, the amount spent and amount taken in will become a *wash* or equal out on the books.

Repairs and Maintenance Expense:

Costs to repair service facility equipment, building mechanicals, company vehicles, and so on, to keep them functional and avoid downtime are deductible expenses. To reduce repair costs and downtime, periodic maintenance on equipment should be conducted during slow periods.

Repair Manuals:

The costs for keeping information systems and repair manuals up-to-date are expensive but necessary.

Security:

This line item is for the monthly and maintenance charges for an alarm system tied to a 24-hour monitoring service and for other related costs such as changing keys of the locks to the building after an employee leaves.

Shipping Expenses:

A common expense is to purchase parts and then have to pay to have them shipped to the service facility (see Freight Charges to Ship Parts for items shipped to the customer). Shipping parts to the facility is really a cost of sales and can be handled in one of three ways.

1. The shipping of a single part can be added to the "cost of parts." The cost is recovered when the item is sold.
2. The shipping (when used as a pass-through cost or when a different markup is desired) can be treated as a separate item and marked up as a cost of sale—other. When sold, it is recorded as sales—other with the shipping charge added to the customers invoice as a separate item. This will help to prevent loss if there is a chance the part will be returned and only the price of the part is to be refunded.
3. When many items (such as small hardware, nuts, and bolts) are shipped into the facility at the same time and it is difficult to attach a charge to a particular item, then it is best to expense the shipping charge for the period under Shipping Expense.

Supplies Expense—Shop:

Shop supplies are items that are not chargeable to a repair, such as spray lubricants, shop rags, grease, oil dry, and cleaners, among other items. Every service facility must have a method to recover these costs either in higher labor rates, a flat fee or a fee based on a percentage of the gross sale.

Tax Expense:

This tax category is for business taxes charged by the state or local government.

Real Estate Taxes:

Real estate taxes are property taxes charged by a local, county, and state governments.

PA Capital Stock Tax:

This line item was used to record the payments on the state tax on Pennsylvania corporations.

Business Privilege Tax:

This is a flat fee charged to all businesses that conduct business in a city.

Telephone Expense:

This expense includes the charges for telephone services and long-distance calls as well as all taxes on the invoice. Facilities must make sure there are phone lines for the manager, the service consultant, the parts specialist, owner, fax, and possibly the computers (a minimum of four or five lines are needed for a small- to medium-size shop).

Tools Expense:

Tools must be purchased by a facility (typically, this does not include common hand tools bought by the technician) for special jobs. These are considered expense items if the cost of the item is not large enough for it to be considered a capital asset that must be depreciated.

Training:

As discussed in Part II of the text on personnel development, technicians must be kept up-to-date. The cost of development programs is a training expense.

Travel:

When an employee travels for business purposes, this account is used for the cost of commercial travel (if the employee drives, the cost is charged as an auto expense), motel/hotel rooms, and other expenses (such as telephone calls, registration fees, materials, and so on). Meals are charged to the meals and entertainment account.

TV Cable:

Charges for a TV cable are necessary for the customer waiting room and, more recently, for e-mail, a Web site, and Internet services needed to keep files for technical support up to date.

Uniforms:

Uniforms, whether rented or purchased, are charged as an expense. The costs to keep them clean are charged to the laundry expense account.

Waste Removal:

The cost for waste removal that complies with government regulations is an important expense that must be documented. A company that is insured to remove the waste and operates in accordance with state and federal standards (properly licensed) must be contracted for waste removal. Managers also should check their credentials and make a check of the waste materials before it leaves the service facility. If a company removes waste and is not properly insured or does not have the proper state permits, the service facility can be held responsible for the cleanup costs incurred by an accident or illegal dumping.

Water and Sewage:

This account represents charges for public water and sewage service.

Other Expenses:

Although the list of given expenses is essentially a list that covers all possibilities, there is always an expense that does not fit any line item account. When this happens, then the Other Expenses line item is used.

Gain or Loss on Capital Equipment:

When used equipment is sold, a gain or loss must be recognized. If the amount received for the equipment is greater than its value on the books (the value less the amount of depreciation or the salvage value if it has been totally depreciated), a gain must be recorded (this will increase the company's taxable profit or generate a capital gains tax). If the equipment is sold for an amount that is less than its value, a loss is recorded (this reduces the taxable profit of the company).

Balance Sheet

The balance sheet is divided into three sections to report what a company owns (assets), what the company owes (liabilities), and how much the company is worth (equity). A typical balance sheet for Bob's Auto Repair is presented in Figure 10-6. Under current assets, the balance sheet reports accounts receivable (money owed to Bob) amounts of $730. In this report, investments are highlighted to show that Bob holds an 18-month CD, and under property, plant, and equipment an automobile has been depreciated to a *book value* of $12,500 and equipment to $20,000. Bob's business assets total $39,093.

Bob's balance sheet liability section shows that the business owes $1,000 in accounts payable (probably to a supplier), $1,400 in taxes (possibly real estate taxes, sales tax, business taxes, or some combination thereof), $10,000 in a note (possibly to a relative or bank for the cash needed to start the business), and $500 in salary (Bob probably owes himself some money because February ended in the middle of the week). When the company assets are subtracted from the liabilities, Bob's equity (what the business is worth but not necessarily its value) is $26,193.

Types of Assets

Assets are basically everything that the company owns. This includes cash, furniture, tools, and equipment, among other items discussed later. Assets can be categorized as current assets and long-term assets. A current assets is a right or property of the business that has

Bob's Auto Repair
Balance Sheet
At February 28, 2005

ASSETS			
Current assets:			
Cash		$1,823	
Accounts receiveable		730	
Office supplies		40	
Total current assets			$2,593
Investments			
Long-term Investments: 18-month CD			4,000
Property, plant, and equipment			
Automobiles	$15,000		
Less: Accumulated Depreciation	2,500	12,500	
Equipment	$25,000		
Less: Accumulated Depreciation	5,000	20,000	
Total property, plant, and equipment			32,500
Total assets			**$39,093**
LIABILITIES			
Current liabilities:			
Accounts payable		1,000	
Taxes payable		1,400	
Notes Payable		10,000	
Salaries and wages payable		500	
Total liablilities			**$12,900**
OWNER'S EQUITY			
Total owner's equity			**26,193**
Total liabilities and owner s equity			**$39,093**

FIGURE 10-6 Example of a balance sheet.

value and can be turned into cash or used up within one year. *Turned into* cash specifically means the asset will either be consumed (such as supplies) or sold (such as inventory) within one year. Types of current assets include:

- Cash is the most common and desirable asset because it can be used to settle liabilities (pay those who are owed money).
- Accounts and Other Receivables is the right to collect the money owed (assumes the debt or those who owe the company money will pay and are considered good).
- Cash Equivalent is excess operational money that is put it into short-term investments for 90 days or less (excess money that will not be needed for at least 90 days).
- Marketable securities are short-term investments or temporary investments that take excess operational money and put it into short-term

investments for 91 days to 1 year (excess money that will not be needed for up to 1 year).
- Prepaid expenses are items such as supplies, insurance, prepaid rent, and so on that the business will use in the future. In practice, as the business approaches the month when the prepaid expense is to be used, the expense will be recorded on the income statement (such as rent) but a check will not be needed because the expense was already paid.
- Merchandise inventory are goods the business owns and eventually sells to customers. When the inventory is sold, it becomes income.

Types of fixed or long-term assets include:

- Long-term investments are for more than one year and can include stocks or bonds of another company as well as land the business plans to resell in the future.
- Property, plant, and equipment (also called fixed assets) that are used in the business and have a useful live of more than one year.
- Land that is used in the operation of the business as well as land improvements (paving, fences, shrubs, lawns, and buildings as well as furnishings).
- Natural resources: timber, mining, oil (cut, mined, or pumped).

Intangible assets include the right to exclusively use something that has value to your business in the future. An intangible asset will help the business earn money, for example, a patent, copyright, or trademark. The cost of the intangible asset can be amortized (similar to depreciation, discussed later) over a period of time so the cost (such as the attorney fees to obtain the patent) can be expensed and recovered over time.

Types of Liabilities

Liabilities also are divided into two groups: current liabilities and long-term liabilities. Current liabilities (see Figure 10-6) are debts that are to be paid off within a 12-month period. Liabilities are listed according to the length of time they are due. For example, an accounts payable liability is the amount owed to other businesses, such as auto parts supply stores, and payments are typically due in 30 days or, at the most, 90 days. A note payable liability is a loan that is usually due within one year. Other examples of liability accounts would be: salaries payable for money owed to employees but not paid, taxes payable to account for taxes to be paid, and unearned revenue that represents when a service or product has been sold and the money has been collected but the product or service has not been delivered. Examples of long-term liabilities include a bank note payable for a loan that does not have to be paid off within the current year, and mortgage payable and bonds payable accounts that may not be due for up to 30 years.

Equity Reports Section

The owner's or stockholder's equity section presents either the owner's investment in the company shown in an owner's equity account (see Figure 10-6) or the amount of stock sold under stockholder's equity. When money is retained by a corporation, the amount is also shown in the stockholder's equity section. A full report of the net income, dividends paid, and amount retained is shown in a *Statement of Retained Earnings* (shown in Figure 10-7).

Note that as per mathematical equation described above:

Equity = Assets − Liabilities.

When equity is a negative number, the business is bankrupt or will be bankrupt if the liabilities (loans) become due before the assets (cash) become available to pay them.

Reports of Accumulated Depreciation—A Contra Account

On the balance sheet, buildings, furnishings, land improvements (land cannot be depreciated), property, plant, and equipment are depreciated, and the amount of the accumulated depreciation is shown in the balance sheet (see Figure 10-6). As depreciation is accumulated, the value of the asset on the balance sheet decreases until it reaches the salvage value (the salvage value can be 0, but also can be an arbitrary number representing the amount of money the item could bring if sold for used parts or scrap metal). After the asset is fully depreciated, it remains on the balance sheet at its salvage value until it is sold or discarded. If it is sold for more than the salvage value, the money made (or lost if it is sold for less than the salvage value) will appear on the income statement as Gain/Loss on Sale of Assets.

Therefore, as depreciation expense is charged each year to recognize the cost of the use of the asset, the amount of the depreciation expense is added to accumulated previous year's expenses to adjust the Accumulated Depreciation on the balance sheet. The accumulated

STATEMENT OF RETAINED EARNINGS
FOR THE PERIOD ENDING DECEMBER 31, 2005

Beginning balance, retained earnings	205,000
Net Income for the year ending	53,825
Subtotal	258,825
Dividends paid during period	(27,730)
Ending balance, retained earnings	231,095

FIGURE 10-7 Statement of retained earnings.

depreciation is known as a *contra account* because it is subtracted from the purchase price of the asset. The revised value of the asset is known as the *book value*. The book value is the amount added to the other assets on the balance sheet to obtain the total assets. Figure 10-6 shows examples of an automobile and equipment that are depreciated on the balance sheet.

To illustrate how depreciation is calculated, an example using the straight line method is presented. In this example, assume a $22,000 truck was purchased and assumed to have a useful life of seven years with a salvage value at of $1,000 at the end of the seven years. To calculate the depreciation expense taken each year, the purchase price would be subtracted from the salvage value and divided by the number of years ($22,000 − $1,000 = $21,000 / 7 = $3,000). Each year, a new book value would be shown on the balance sheet. After the first year, the book value would be $19,000 ($22,000 − $3,000 = $19,000) and after the second year another $3,000 would be added to the accumulated depreciation to give a book value of $16,000 for the truck.

Therefore, each year $3,000 would be charged to depreciation expense in the income statement ($750 on each quarter) to reduce the net income and lower the tax liability of the owners. On the balance sheet, each year the $3,000 for depreciation would reduce the value of the truck until a $1,000 balance remains. When the truck is sold or disposed of, it and the $1,000 would be removed from the balance sheet (if sold for more than $1,000, the business would owe a capital gain tax; if sold for less than $1,000, the loss would reduce the taxable income).

Although the straight line method may be used internally by companies to present a record of their assets, it is not an acceptable method under current IRS code. For IRS tax reporting, the acceptable methods an accountant may use are the

- Accelerated Cost Recovery System (ACRS used for assets purchased from 1981 to 1986)
- Modified Accelerated Cost Recover System (MACRS used for assets purchased from 1987 to the present)

The ACRS or MACRS permits companies to fully depreciate their assets (the salvage value is 0) in a given period of time (depending on the type of asset) for percentage amounts found in the IRS schedule. This permits companies to replace their equipment in a shorter time period in order to benefit the economy.

Accrual versus Cash Basis Accounting

Businesses may use either the **cash basis** or **accrual basis** form of accounting.

In the cash basis method:

- A sale is recorded when cash (payment) is received.
- Expenses are entered when cash is paid.

In the accrual method:

- A sale is entered at the time the sale is made (whether or not cash is received).
- Expenses are entered at the time of the purchase (not when cash is paid).

Therefore, if Bob's Auto Repair used the cash basis method and fixed a car in February but the invoice was not paid until March, the sale would appear in the income statement for March, not February. If Bob's Auto Repair used the accrual method, the sale would be recognized in the income statement for February.

Corporations are required to use the accrual method, whereas proprietorships and partnerships may use the cash method. The flow diagram for transactions shown earlier would be modified as shown in Figure 10-8. Notice that both methods still use an income statement and a balance sheet. However, the method determines when the transaction is recognized and recorded on the accounting statements.

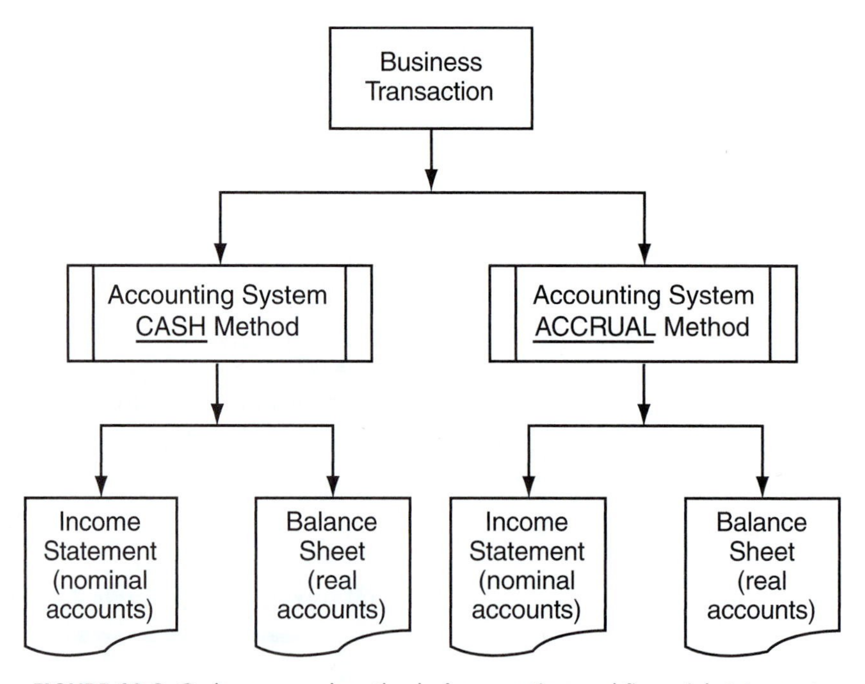

FIGURE 10-8 Cash or accrual method of accounting and financial statements.

Final Thoughts

The accounting system used by a service facility must fit the size and type of business being operated. For a large corporation, a complex double-entry accrual accounting system with numerous expense accounts to keep track of charges is required to comply with the GAAP (Generally Accepted Accounting Principles) and meet the needs of managers, owners, and the IRS. A smaller proprietorship may use a less formal system of money in–money out cash method accounting with a limited number of expense accounts that does not require much more than taking care of a personal checking account. Regardless of the method and size of the system, all businesses must monitor their financial reports in relation to their strategic objectives with benchmarks through the analysis of their financial reports.

An examination of the income statement is critical, as it offers numerous insights into the effects of and relationships between income and expense line item accounts and profit. An important practice, however, is that the performances of the business and the income/expense relationships being compared must be calculated and recorded every week, month, and quarter. In addition, this information should be placed on a chart or graph to present a picture of performances. Sometimes the illustrations of the ups and downs of different accounts can serve to impress upon people the importance of the changes taking place.

The greatest benefit of the balance sheet is to help managers and owners recognize the effect of the business operations on owner's or stockholder's equity. This measure is an indicator of the health of the business because it is directly impacted by monthly profits or losses. In addition, the value of the assets and the amount of liabilities also must be monitored on a monthly basis. When the balance sheet has negative equity and not much cash on the books, management is in a difficult (almost a state of emergency) position as compared to managers with a positive equity balance that is increasing on a monthly basis along with a healthy cash balance. As a result, every business, even a small business, needs to prepare a balance sheet.

Although an income statement and balance sheet are important to managers, managers also must be familiar with the circumstances that influence the major income and expense accounts. Although all line item expense accounts must be monitored constantly and adjustments must be made to keep them in line, the most important line item accounts to a service facility are those that record the sales of the business and the cost of those sales. The accounts for sales and cost of sales contain the largest amounts recorded into the accounts on the income statement plus they have the most influence on the balances in the equity and cash accounts. For this reason, the final chapters of this book focus on labor rates, markups on merchandise, and analysis methods.

Review Questions

1. What is the difference between real and nominal accounts?
2. Define each of the following: assets, liabilities, and equity accounts.
3. Describe each of the three sections, accounts, and purpose of the income statement.
4. Describe the three parts and purpose of the balance sheet.
5. What is book value?
6. What is the difference between cash basis and accrual basis accounting?
7. What is the purpose of depreciation?
8. What is the purpose of a merchandise inventory worksheet?

CHAPTER 11

PAYING EMPLOYEES AND THE LABOR SALES WORKSHEET

LEARNING OBJECTIVES

Upon reading this chapter, students should be able to:

- *Explain how salary, labor rate, and straight-time methods to pay employees differ.*
- *Describe how the labor rate is used to inform customers of their charges, to calculate labor sales, and to determine how much a technician gets paid.*
- *Explain how warranties use the flat-rate system.*
- *Identify the three sections of the labor rate worksheet.*
- *Calculate total labor sales, total cost of labor, total expenses, and net income on labor sales.*
- *Explain how percentages calculated from the labor sales worksheets can be used to identify specific problems.*

Introduction

Salaries and wages, commonly referred to as the payroll, is the greatest expenditure of money at service facilities. Payroll costs, discussed in this chapter, cover the salaries and wages paid to technicians, service consultants, and managers. As this chapter explains, service managers and owners must be aware of the methods that can be used to pay different personnel so that the best method can be used in their facility. In some cases, a person may be paid for the amount of hours spent at the facility; other methods recognize the employees' skill and intensity of the work; and some pay plans reward employee performances.

As explained in Chapter 5, the difference between exempt and nonexempt employees is a major factor when paying workers. Exempt workers are paid for the job performed and not the number of hours worked. Generally, exempt workers are part of the management team. In contrast, nonexempt employees must be paid overtime if they work over 40 hours. The failure to pay personnel according to the federal labor law can become a costly legal problem, as many employers can testify. The purpose of this chapter is to explain how staff exempt and staff nonexempt employees may be paid. The definitions presented in the last chapter are important; for instance, hourly pay (payment for each hour worked), **flat–rate pay** (payment for each billable or flat-rate hour charged to a customer, commonly used as a pay method for technicians), salaries (pay for a period of time), bonuses (payment when a target is achieved), and commissions (payment based on sales a transaction, typically a fractional amount is paid to the employee who makes the sale) are used as per their definitions.

Paying Exempt Employees

Positions classified by the federal government as exempt are executive, administrative, and professional positions, as well as outside sales personnel. Exempt employees are not paid overtime when they exceed 40 hours per week; however, the law specifies that the minimum amount they are to be paid ($455 as of 2006). The law requires that employers keep records on each position, such as the person's job position and description, pay, hours worked, and so on. This is to prohibit employers from giving a worker a fancy title in order to get around overtime pay.

Exempt employees, may be paid a combination of a salary, bonus, and commissions, although the federal minimum must still be met. A salary may be paid for a period that is mutually agreeable between the employer and employee, such as a week, two weeks, and so on. A bonus payment maybe added to a salary when a level of performance, such as sales or production, is exceeded. For instance, an employee may be paid a low salary and a bonus when sales exceed a predetermined amount. Or the person may receive a graduated bonus schedule so that he/she is motivated to increase his/her income by striving for higher and higher performances.

Commissions, as opposed to a bonus, are based on sales, such as a 5 percent commission on gross sales (parts and labor) or a commission on parts only. In some cases, a service consultant's earnings might depend primarily on his/her commission with a salary guarantee or vice versa. The use of a commission payment has a number of pros and cons. Although a service manager might wish to motivate a service consultant (management position) to promote sales, the manager would not want the person to pressure customers into repairs or sell unnecessary parts for the sake of earning a higher commission.

Paying Nonexempt Employees

Technicians and other nonmanagement employees are nonexempt employees as per the definition of the federal government in the Fair Labor Standards Act (see Chapter 5 for definitions of exempt and nonexempt). These employees must be paid the federal or state minimum hourly wage (whichever is higher) and overtime (one and a half times the employee's hourly wage) when the employee works over 40 hours per week. This does not prohibit them from being paid a bonus or commission or under a flat-rate system. When a flat-rate or a commission is used to pay nonexempt employees, however, the employee must earn at least the federal minimum hourly wage each week and receive overtime pay if they are required to work over 40 hours. Failure to pay a staff nonexempt employee at least the federal minimum hourly wage (regardless of the pay plan used) puts the employer at odds with federal and state labor laws.

Collective Bargaining Units

Some businesses have employees (such as the service department technicians) that are part of a collective bargaining unit. The purpose of a collective bargaining unit is for a group of employees (called union members) at a service facility to negotiate their wages, benefits, and working conditions as a group with the owners/management of the service facility. The agreement reached between owners/management and the collective bargaining unit is recorded in a legal document called a union contract, as shown in Figure 11-1.

To help the collective bargaining unit negotiate and enforce contractual agreements with the owners, an organization, referred to as a union, is used. Unions are large organizations that represent many different collective bargaining units. Well-known unions, such as the Teamsters and

FIGURE 11-1 Technician's copy of a union labor agreement.

AFL-CIO, are large national unions that represent thousands of members at many businesses. In return for representation, the union collects dues from the members and often oversees their retirement accounts (the pension fund). After negotiations are concluded between the owners and the union representatives, a contract is signed and enforced by the union. Employees obtain legal representation from the union when the contract is violated. The contract is then renegotiated periodically (typically, every few years) to improve wages, benefits, and working conditions.

In order for a union to get started at a business, the workers must vote to have a collective bargaining unit that is represented by a specific union, such as the Teamsters. There are federal and state laws that govern this process as well as how a union/collective bargaining unit and employer can interact with each other; however, these are beyond the scope of this book. The contract will include topics such as:

- Technician pay (and anyone else included in the bargaining unit such as technician helpers, parts specialists, and possibly service consultants depending on the limitation of their duties) as per wages negotiated in the union contract.
- Discipline (reprimand, suspend, terminate) of employees by following specific steps found in the union contract.
- Grievances (complaints) filed by employees about their treatment.
- Working arrangements between management and union stewards, who serve as a liaison between the business, the union and members (employees); for example, correspondence with the union goes through the union steward.

In order for managers to work effectively with unionized employees, they must carefully read and understand the contract between the owners and the union. Many managers find that working with a union is not difficult and can even make their job easier because it takes away the burden of creating some operational procedures needed in a shop. For example, if an employee wants a pay raise, the manager simply refers the employee to union contract. When an employee chooses not to show up for work, union contract procedures are followed (which typically involves documentation, submission of evidence, and specifies how many times this can happen before the employee is terminated).

The Flat-Rate System

Under the flat-rate system, a technician is paid for every hour of work that can be billed to the customer. Regardless of the number of hours a person works, he/she is paid for the *billable hours* generated (flat-rate hours generated). Technicians paid on an hourly basis are paid for each hour they are at the facility in a working capacity even if they are not working on an automobile or performing some other assignment.

The number of flat-rate hours allotted to a job comes from a labor time manual or computer database. To obtain the times, each automobile manufacturer conducts time and motion studies to determine how much time each repair will take on the automobiles they manufacture. The labor times not only include the time the repair takes but also are inflated to include time to obtain parts from the parts department as well as other normal events (such as cleaning parts) that occur during the repair process. These times are then published in the manufacturer's labor time guides (or computer database).

New-automobile dealers are required to use flat-rate times when they perform warranty work for the manufacturer. When a warranty job comes into a facility, the service consultant must look up the labor time for the job and post it to the repair order. Provided that there are not any unusual circumstances that would require additional labor time (such as a burned wiring harness that must be repaired/replaced, bolts that are broken, etc.), the job is billed to the manufacturer at the specified labor hours multiplied by the labor rate that the manufacturer has approved for the dealership. The labor rate approved by the manufacturer for the dealership is based on several factors (such as location of the dealership, among others). For example, assume the replacement of a head gasket under the (fictitious) Alpha Motor Company warranty pays five hours as per the Alpha Motor Company labor time guide. Also assume the new-automobile dealership that performs the repair has an Alpha Motor Company approved warranty labor rate of $50 per hour. This means the charge to Alpha Motor Company is $250. The payment to the new-automobile dealer is typically made through a credit for the purchase of parts and not by a check sent from Alpha to the new-car dealer's service department.

Naturally, there are repairs in which additions are made to the labor times allotted for a job, such as a repair that requires the discharge and recharge of the air-conditioning unit. Additional time also may be added to a repair when the technician experiences problems in the repair process, such as broken head bolts that must be extracted. The time added for these additional repairs are typically paid straight time, which is the amount of time it takes to make the repair rather than a predetermined amount of time found in the labor time manual.

When a job is not paid under a manufacturer's warranty contract, many (but not all) new-automobile dealers and most independent repair shops base their labor times on an aftermarket book, such as Chilton or Mitchell. The Chilton and Mitchell guides and computer databases list manufacturer labor times for all of the jobs that may be performed. A service consultant or manager will find that the Chilton and Mitchell guides have times that are slightly higher than the manufacturer's warranty labor guides. This is often because as automobiles age, repairs usually require more time for cleaning, and the bolts and fasteners are likely to be more difficult to remove because they are rusted into place (although not typically stripped or broken, which, if they are, may require additional time be allotted for the repair).

Unlike a manufacturer's warranty repair, a customer is given an estimate before work begins. To give an accurate estimate, the labor time is looked up and multiplied by the labor rate of the service facility and, as suggested in the authors' previous book, a cushion is often added to help cover any unexpected problems. When the job is finished, the cushion should ensure the final invoice is less than the estimate given to the customer. When an unexpected problem occurs and the final invoice will be greater than the estimated amount (with the cushion included), the customer must be notified as soon as possible so that approval can be obtained for the additional charges.

A Customer's Invoice and the Flat-Rate System

Unlike a manufacturer's warranty job, many customers do not understand that the time it actually takes a technician to perform the repair is not related to the amount of time they are charged for the job. An experienced technician may take less time to do a job than an inexperienced technician. In either case, the customer pays the same amount of money regardless of how long it took to do the job. In fact, many state laws require service facilities to disclose to customers how the charges on their invoice are calculated so that customers understand when they are billed a flat fee regardless of the time it takes to do the work.

To illustrate how this works, assume a customer is told that the labor charges for the repair will be the number of hours allotted for the job multiplied by the labor rate. The service consultant finds that the Chilton Labor Time guide allots 1.5 hours to do the work and this is multiplied by the service facilities labor rate of $50 per hour. The labor charge to the customer is estimated to be $75 (without any cushion added into the job), regardless of the number of hours the technician must take to fix the car.

Flat-Rate Pay Earned by Technicians

While the preceding illustration exhibits the cost to the customer and total amount received by the facility, the manager also must recognize the cost to provide the labor. On the cost side of this job, assume the technician is paid $16 per flat-rate hour (including Social Security and other benefits). For this job, the labor for the repair will be $24 (1.5 × $16 = $24). When the technician is paid by the flat-rate system, he/she will be paid 1.5 hours, whether the job takes him 1.25 hours or 2 hours. For comparison purposes, when the technician is paid by the hour, the labor paid for the repair is based on the time it actually takes to perform the work (1.25 or 2 hours) and not the number of flat-rate hours (1.5 hours) billed to the customer.

The flat-rate technician's pay at the end of the week is based on the total number of flat-rate hours allotted to him/her for repairs times accumulated during the week. Their final pay is then based on the total of flat–rate hours earned for the week (not the time it actually took to complete the jobs). For example, if a technician accumulated 50 flat-rate hours for repair jobs for the week at $15 per flat-rate hour, his/her paycheck at the end of the week would be $750, regardless of the number of hours worked.

For a manager, if the facility charged $50 per flat-rate hour, it would receive $2,500 from the customers (50 hours × $50 labor rate = $2,500) for the work performed by the technician. The difference between the wages earned by the technician and the amount earned by the facility must help to pay the salaries of the service consultant and managers, line item expenses (Chapter 9), and the return on the investment of the owners.

Flat-Rate Technician Wage Guarantees

The automotive repair business is very cyclical with seasonal and weekly ups and downs in the volume of business conducted by the facility. A facility technically would commit an illegal act if it paid a staff non-exempt employee, such as a flat-rate technician who works 40 hours per week, less than hourly minimum wage. This could happen when, for example, a $10 per flat-rate hour technician earned 15 hours for the week and was paid $150 but worked 40 hours. This would clearly pay the technician less than the federal minimum hourly wage and put the employer at odds with a government bureau if a complaint were issued. To avoid this problem, some service facilities institute a guarantee that adds time to a flat-rate technician's flat-rate time to ensure he/she at least meets the minimum wage standard. For example, a $10 per hour flat-rate technician may be paid a guarantee of 23 flat-rate hours to ensure the minimum wage is paid. If a difference of six flat-rate hours was paid (17 hours earned and 23 hours paid), the six hours would be called unapplied time. Depending on the employer, the guarantee could be set at any number or could be in the form of a minimum dollar amount or flat-rate hours added to what was earned. Unionized technicians often have a 100 percent flat-rate guarantee, which means they are paid for 40 hours or the number of flat-rate hours earned, whichever produces the highest wage.

When a technician has prolonged periods of unapplied time (when weekly flat-rate hours earned are less than the hours worked) company policy may dictate that the employee should have their job performance and pay reviewed. In some cases, a technician may be taken off flat-rate and paid hourly at a corresponding lower hourly rate. In some cases, company policy may have provisions that disciplinary action, such as dismissal, may be taken when the employee cannot improve performances to a minimum level after a period of time for correction. As discussed later, managers must be aware that, in some cases, a failure to obtain minimum expectations may be because of work assignments or because business is slow. This is often the case in some smaller cities and towns with seasonal populations, such as college towns, where students leave for the summer, tourist areas, and factory towns that shut down for retooling. In these situations, owners may elect to lay off employees, who can then collect unemployment until the business conditons improve and they can return to work.

Flat Rate and Job Assignments

For technicians paid under the flat-rate method, the work assigned to them is important. When a job takes longer than the time allowed under the flat-rate time provided (for example, a brake job on a car that is old with rusty parts), then the technician will lose time and ultimately money. Although technicians expect to lose time on some jobs, when it becomes chronic, then morale problems are created. In some cases, technicians have quit their job because the work assignments were not fairly distributed. Consequently, service managers must monitor work assignments and compare flat-rate hours earned by the technicians against their individual targets. It is management's responsibility to make sure work is being handed out fairly so that each technician (hourly and flat-rate) is meeting his/her flat-rate hour target (also called a **flat-rate objective**). This is important because when a technician does not meet his or her target, not only does this person lose money but also the service facility does not make as much money as it should. Therefore, managers must monitor what and how work is distributed. It is not unusual for the person assigning repair jobs (team leader or service consultant) to hand them out to people based on his/her personal feelings toward a person or to give easier jobs to technicians who are slower, older, have greater family responsibilities, and so on.

Use of Time Clocks on Flat-Rate Jobs

To monitor the time allotted and time taken to complete jobs, the authors used a time clock. When a work order was prepared, the allotted time from a labor guide for the repair was written on it. When the technician began to work on the automobile, the technician punched the time on the technician's copy of repair order (called the **hard copy**) by inserting it into the time clock. When the work was completed, the technician inserted the repair order into the time clock again to punch in the time. The actual time taken for the repair was then calculated and recorded on the technician's repair order hard copy. At many modern shops, computers have replaced paper repair orders. In these setups, a technician logs onto the computer when the job is started and again when it is finished. In either case, this practice collects very accurate data about how long it takes to do a job in comparison to the flat-rate hours charged to the customer. This is the basis to calculate the technician's flat-rate hour efficiency, covered in detail in Chapter 15.

Paying by Hourly Rate

As noted earlier, when employees are paid on an hourly wage rate (or straight time, which is paid to flat-rate technicians when they encounter problems on a job or if the service facility runs out of work), they are paid for the time they are at work regardless of their activities. Although some employees are paid by the hour, most service facilities use the flat-rate system to pay technicians (particularly highly skilled technicians). Those who assist technicians or are lower-skilled technicians who perform maintenance services (oil changes) are paid by the hour. Unskilled employees who might perform custodial or clerical duties typically are paid by the hour.

A service manager has two concerns with respect to paying hourly personnel: one is the rate of pay and the second is keeping an accurate record of the hours worked. With respect to the amount to be paid, the rates of similar jobs at similar service facilities should be obtained. The amounts paid also should be compared to the pay rates of other employees at the facility to ensure the wage amounts are reasonable and in alignment. For example, if a loyal unskilled employee (car washer or parts runner) has worked at a facility for a number of years and has received minimal increases each year, the person may eventually be paid more than an amount appropriate for the job. This can cause a wide variety of problems for the business. This is why unskilled workers may be awarded a bonus without an increase or a very small increase in pay, such as an amount equal to the inflation index.

Keeping track of hourly or wage payroll personnel is an undesirable task, as anyone who has done it can testify. One approach may be to have the employee check-in with another person when reporting to work. This method is not desired because it disrupts the work of the employee who checks in the workers. The other option is to use a time clock or computer login. When the worker reports to work or leaves work, the person must take their time card and punch it. If a person forgets to punch in, it is their problem and they don't get paid. The supervisors, however, must make sure each employee punches out. When an employee does not punch out, the manager must assume the employee stopped working at the normal time or worked until their last scheduled break.

A common misconception about wage payroll is that if an employee hangs around and does not punch out, he/she can still get paid and even collect overtime. As a result, at the end of the workday, some employees must be monitored and told to punch out. In addition, company policy should be clear that management must approve all overtime in advance. Employers must understand that the labor law stipulates that when a person works over 40 hours within the workweek they must be paid overtime. As a result, managers must keep an eye on the amount of time a wage payroll employee works so that they are dismissed for the week when they reach 40 hours, otherwise overtime must be paid.

The Labor Sales Worksheet

A manager must monitor costs and expenses constantly so that he/she can make timely decisions. The manager is like a coach who must monitor the game constantly to make decisions. Imagine a coach who does not prepare a team for a game or does not even watch the game! To make timely decisions, the manager must collect data information and make observations in order to be aware of situations in which advantages are offered and in which potential problems are evolving. The information collected on labor sales and costs are placed in a worksheet that is similar to an income statement but that is limited to information relevant to the operation being managed.

A **labor sales worksheet** designed for the recording and analysis of labor sales is shown in Figure 11-2. This worksheet presents information on labor sales, cost of labor sales, and related expenses. A similar worksheet would be prepared for an examination of parts sales and related costs and expenses. These worksheets are common and referred to as *managerial accounting worksheets.*

The labor sales worksheet presents multiple income sources. The first line item, *gross labor sales,* is usually the largest source. The second line item is for *transfers from the parts sale area* and is only used when a parts supplier reimburses a service facility for labor to install defective parts. The third line item is for *discounted labor sales* and indicates when owners or the manager approves the use of a lower labor rate than posted. Although the labor rate should never be reduced, there are situations, such as fleet contracts, when a lower labor rate may be necessary.

The next section in the worksheet exhibits the cost of labor. Specifically, the technician's wages are shown as well as the costs associated with the employer's share of Social Security and Medicare taxes. To calculate the technician's wages, the technician's flat-rate hours or hours worked (for technicians who are paid by the hour) are entered into the worksheet from the service manager's records, which should include the hours worked and flat-rate hours (in weekly terms) for each technician.

Although the information in the manager's records may be weekly, this managerial accounting worksheet is set up for monthly reviews. Therefore, weekly line items (such as technician's hours) must be converted to monthly line items. To do this, the conversion factor is 4.3 (some may think it should be 4 since there are 4 weeks in a month, but there are 52 weeks in a year which when divided by 12 = 4.3). For the flat-rate technician in Figure 11-2, the wages are calculated each month on an average of 50 flat-rate hours produced each week. This equates to 215 flat-rate hours (50 WFRH × 4.3 = 215 MFRH). For the hourly technician, who works 40 hours per week, this equates to 172 hours per month. When overtime is paid each week (for technicians who are paid by the hour), the number of hours of overtime is multiplied by 1.5, then added to 40 and multiplied by 4.3. For example, 2 hours of overtime each week (1.5 × 2 = 3) is 3 hours added to 40 hours to equal 43 hours multiplied by 4.3 is 185 hours (this is rounded and is included in the worksheet because it is used for payroll purposes).

After the labor sales and cost of labor are determined, the manager should calculate the gross profit and related percentages by applying the following equations:

Total Labor Sales − Cost of Labor = Gross Profit (see Figure 11-2 = $17,349).

Total Cost of Labor/Total Labor Sales = labor cost as a percentage of labor sales (see Figure 11-2 = 38.48%).

Gross Profit/Total Labor Sales = gross profit as a percentage of total labor sales (see Figure 11-2 = 61.52%).

INCOME LABOR SALES AREA			**Receipts**	**Percentage**	**Target Metric**
Gross Labor Sales			$27,750	98.40%	
Transfer from Parts Sales Area			$0	0.00%	
Discounted Labor Sales			$450	1.60%	
Total Labor Sales			$28,200	100.00%	100%
COST OF LABOR					
Flat-Rate Technicians	**hours/ month**	**Flat-rate wage**			
#1	220	15	$3,300	0.00%	
#2	190	13.5	$2,565	0.00%	
#3	175	13	$2,275	0.00%	
			$8,140	0.00%	
Hourly Technicians	**hours/ month**	**Wage per hour**			
#1	172	11.25	$1,935	0.00%	
Total Hourly Technician Wages			$1,935	0.00%	
Total Technician Wages			$10,075	35.73%	
Social Security/Medicare			$776	2.75%	7.7% of tech
Total Cost of Labor			$10,851	38.48%	<40%
Gross Profit			$17,349	61.52%	>60%
EXPENSES					
Support Staff					
Service Consultant Wages			$0	0.00%	7% gross sales
Service Manager Wages			$2,350	8.33%	9% gross sales
Commissions/Bonus			$600	2.13%	optional
Social Security/Medicare Tax			$227	0.81%	7.7% staff wages
Employee Benefits					
Uniforms			$270	0.96%	$50/employee/mo
Retirement			$487	1.73%	optional
Medical Insurance			$2,200	7.80%	optional
Development Programs			$0	0.00%	optional
Operations					
Insurance			$1,000	3.55%	
$200/employee/mo					
Equipment Leases					
1			$125	0.44%	optional
2			$65	0.23%	optional
Telephone Expense			$202	0.72%	<2%
Advertising with Phone Book			$450	1.60%	>3%
Tool Expense			$45	0.16%	actual
Garbage Removal			$120	0.43%	actual
Facilities					
Rent			$2,100	7.45%	0.4 to 0.6/sq ft/mo
Property Maintenance			$0	0.00%	0.2 to 0.6/sq ft/mo
Property Tax			$0	0.00%	.05 to .08/sqft/mo
Heat / Light / Electric			$175	0.62%	.12 to .2/sqft/mo

FIGURE 11-2 Labor sales worksheet.

INCOME LABOR SALES AREA	Receipts	Percentage	Target Metric
Administrative			
Bank Charges	$45	0.16%	1–2% of gross sales
Office Supplies	$0	0.00%	1–2% of gross sales
Information Management	$0	0.00%	1–2% of gross sales
Other Expenses	$38	0.13%	1–2% of gross sales
Total Expenses	$10,499	37.23%	<45%
NET INCOME	$6,850	24.29%	>15%

FIGURE 11-2 *continued*

Target Metrics

To help understand whether the numbers on a management worksheet are reasonable, target percentages are provided for each expense category and many line items. In the model, gross profit (greater than 60 percent), expenses (45 percent or less), and net income (18 percent or greater) would be targets. Within the expense category, each line item would have a target metric. A target in the analysis typically assumes a percentage, which is basically a ratio, used in accounting for analysis purposes.

Some targets in an analysis, however, may be given as units, dollars, or derived from other information so that a dollar-based target can be established. To ensure the user knows that the target may be in a unit other than percent, such as (the number of gallons of heating oil used), the term **target metric** is used. To evaluate a specific line item, it is compared to a target metric. For instance, a target metric can help determine if an expense amount is too great. The target metric also can determine whether an entire category of expenses is too high relative to the labor sales (it may not be that the expense category is too great as much as the income is too low).

Therefore, the inclusion of a target metric for each expense line item allows the worksheet to be used as an expense-driven analysis. This is useful when a prediction about a business expansion sales target is needed. For example, if expenses are estimated to be $32,000 and they are to be 45 percent of sales, the sales target would be $71,111 (32,000/.45 = $71,111). This means before starting a new business, a model could provide some useful quantitative information for budgetary purposes (this will be discussed in Chapter 12).

The targets used in Figure 11-2 are based on information extracted from business records, compiled, in part, from government statistics for *profitable* automobile repair facilities. It should be noted that the percentages, such as gross profit margin for smaller businesses, is not the same for larger businesses. Therefore, the example presented should be considered a suggested model or learning tool to be modified based on business size, location, and specific needs (some expenses relevant to a certain type of business may not appear on this worksheet). Although the model target metrics may be considered realistic, they are not exact for any given service facility type, size, or location. This means the best data is actual expense data recorded in the income statement. When creating a model, the plan should be to use the actual billable expenses (not tax write-off expenses such as depreciation expense) so that the manager ultimately can ensure a service facility has enough gross profit margin to pay its expenses and obtain enough net income for the owners.

Monthly Expenses

The expenses for each month must be entered into the worksheet in much the same way as the cost of labor. In the expense category, there are several expense subcategories of expenses (see Figure 11-2). They are as follows:

- Support Staff
- Employee Benefits
- Operations
- Facilities
- Administrative

As shown in Figure 11-2, within each subcategory are the line item expenses. When possible, actual expense data should be used to gain greater depth in an analysis; however, there are instances in which expense information is not available, known, or provided to the service manager. In this case, the target metric column is useful because it contains a theoretical range of where the expense should fall so that an amount can be estimated. Conservatism would dictate that the higher side of the range should be used for analysis purposes. This would mean that if the rent was unknown but the service manager knew the service facility was 4,000 square feet, the rent could be estimated. For example, if the facility used 3,500 sq. ft. at $.60 per sq. ft. (the high end of the target metric), the expense allocation would be $2,100. This may not be exact, but for purposes of analysis it is probably close enough (depending on where the facility was located).

Labor Sales Area Net Income

After the expenses are calculated and totaled on the worksheet, they are subtracted from the gross profit to present the net income. Both the expenses and net income percentages of the gross profit should then be calculated as follows:

Total Expenses/Gross Profit = Expenses as a percentage of Gross Profit (see Figure 11-2 = 37.23%, which is less than 45%).

Net Income/Gross Profit = Net Income as a percentage of Gross Profit (see Figure 11-2 = 24.29%, which is greater than 15%).

Analysis of Labor Sales Area Data

The analysis of the data compares the percentages at the data relative to the target metric shown in the column on the right side. After the worksheet calculations, the gross profit percentage of 61.52 percent would be compared to the target percentage of less than 60 percent. This can tell the service manager whether the sales, costs, or expenses are operating in an optimum range or not. (In Figure 11-2, all calculations are within the optimum range.) Of course, the greatest concern to the manager is that the output (gross profit, expenses, and net income) of all three catagories are within the targeted metric range.

Although there are many ways to look at the data, the objective is to examine the category targets and their components. When they are outside the optimum range, then the manager should ask why. In some cases, a line item expense within a category is too high or sales are too low. In other words, the worksheet is like scan tool data; it tells the manager what is out of range to help him or her to focus on a specific system. Likewise, once the worksheet analysis shows a problem area or areas, the manager, like the technician, must investigate to see what is wrong and determine how it can be corrected.

Expense and Cost of Labor Estimation

When an analysis is performed for expansion purposes or for a business that does not exist (an empty building where a business might go), then expenses must be estimated. Many major expenses can be estimated when actual costs are not available by the method described earlier. Some costs, however, such as employee wages, cannot be estimated using this method, and other research is necessary. For example, a service consultant's wage must be estimated based on the going wage for a given region of the country. In most areas of California, a wage of $25,000 per year would be considered low, whereas in North Dakota this same wage might be viewed as reasonable. Research also would be required for equipment that the business would need. To do this, the service manager would refer to tool catalogs to estimate the cost of tools the business might need over a period of time. Therefore, the worksheet relies on research to generate the information needed to get the accurate costs the service manager can use.

Once the expenses are estimated, the total will reveal the amount of money needed to cover the monthly bills. This number is in most circumstances not to be any greater than 45 percent of the income. Therefore, if the worksheet expenses are $8,000, which is to be 45 percent of sales, then labor sale income must cover the expenses (assuming labor sales is the only income being received). The amount of the income needed would be $17,778 ($8,000/.45 = $17,778). To estimate the amount of money available to pay technicians from an income of \$17,778, the amount is multiplied by 40 percent to equal \$7,111 (\$17,778 × .4 = \$7,111). Naturally, if the model is accurate and reasonable (for example, can technicians be hired for \$7,111?), the projected net income can be calculated, which would be \$2,667 (\$17,778 − \$7,111− \$8,000 = \$2,667). This amount must be enough to cover the expected returns on the capital investments plus any loans needed to start the business.

Net Income and Cash Flow

Earning an adequate net income is the main objective of any owner; however, whether the net income earned will be enough is beyond the scope of this book. In general, the net income earned should be at least 15 percent of the sales. Naturally if there is not enough income to cover expenses, then there will not be a profit. In some cases, there also can be a profit but not enough cash to pay the bills. This is typically due to several reasons. One is the service facility may be owed money from various sources (this is referred to as accounts receivable) and delayed payments cause bills to come due before cash is received to pay them (called cash flow problems). To illustrate this, assume a fleet has a number of invoices to be paid to a service facility. This causes cash flow problems when the service facility has to pay for parts and labor to fix the fleet vehicles before the fleet's payment is received.

There also may be cash flow problems when the net income cannot pay for the loans the owners borrowed to start the business. Because loans to start or expand the business typically are paid from the net income (especially for proprietorships and partnerships), the failure of the net income to cover the principal and interest (interest is an expense that would appear on a traditional accounting income statement but does not appear on the worksheet) may cause a cash flow problem.

To help owners understand how money flows into and out of their business, a traditional accounting statement called a **Cash flow statement** is useful. Although this is beyond the scope of this book, this statement examines the change in cash for a given a period of time. Although this is helpful information, creating a cash flow statement using the indirect method (derived from the income statement) can be complex. Furthermore, interpreting why the changes occurred requires extensive evaluation of the balance sheet and requires a depth of knowledge about accounting that is beyond what most service managers need to do their job effectively.

Final Thoughts

When discussing payroll, one of the most frequently asked questions is how much a facility should pay employees. This is a tough question, because employees who are overpaid can cause financial problems that can run a facility out of business. At the same time, employees who are underpaid can cause a high turnover of employees and the best ones often leave first. To help retain good employees, a system that rewards performance is common. For service consultants, direct commission on sales is typical. For technicians, the flat-rate system is commonly used. Management can expect that they will not see very good performance when the technicians, as well as the service consultants, are not paid a competitive wage or provided with a good benefit package. Managers and owners can attempt to find out what their competitors pay, but that is difficult because there are so many variables that influence a pay rate. For example, the hourly rate will depend on whether a technician is considered an A-, B-, C-, or D-tech. A service consultant at one facility may be a staff exempt employee, whereas another may be staff nonexempt. The benefits also may vary from facility to facility, and even paid vacation and sick days are likely to be different. Finally, trying to collect information about employee pay and benefits can be impossible because no one is likely to tell the truth. In reality, the comparison is even more difficult because what an employee is paid is basically a function of the business itself and what the service facility charges customers. A poorly run and inefficient operation that cannot charge customers enough cannot expect to pay as well as a respected and efficiently managed operation.

The best approach to setting a pay schedule is by looking at the figures on the labor sales worksheet. The owners and managers must set their target metrics honestly with respect to the percentage they believe their employees should receive from the business. If the business is successful and makes more than expected, the owners and managers must decide if the employees will share in the profits through either commission, bonus, or flat-rate pay. When management and owners are not competitive by paying their people fairly, then high turnover is likely. When owners are too generous, the job security of the employees is at risk. In other words, pay rates are difficult to set. It is doubtful that workers, managers, and owners can be satisfied. The only clear way to determine when there is enough is to prepare a business plan with visionary goals with reasonable and clear financial objectives for wages, raises, then benefits, and set out to achieve them (both as a business and personally).

Review Questions

1. How do salary, labor rate, and straight-time methods to pay employees differ?
2. How is the labor rate used to inform customers of their charges, to calculate labor sales, and to determine how much a technician gets paid?
3. How do warranties use the flat-rate system?
4. What are the three sections of the labor rate worksheet?
5. How are percentages calculated from the labor sales worksheets used to identify specific problems?
6. Explain the purpose of target metrics.

CHAPTER 12

SALE OF PARTS

LEARNING OBJECTIVES:

Upon reading this chapter, students should be able to:

- *Describe the costs incurred by a service facility when selling parts.*
- *Present the different methods to prepare a markup for parts.*
- *Calculate markups.*
- *Explain the advantages of just-in-time parts delivery.*

Introduction

As discussed previously, labor sales provide the greatest income earned at most service facilities. The second greatest income to be earned comes from the sale of parts. To manage the business activities of a service facility accurately, labor sales and parts sales must be analyzed separately. Although this is common at most service facilities, the process may differ depending on the business, such as a new-car dealership where the volume of parts sales at a dealership is large enough that a separate parts department is created. These parts departments are usually run by a separate manager and an extensive inventory of parts and supplies is maintained. In these cases, full-time employees are hired to order, unload, stock, look up, sell, deliver, and maintain the parts inventory to be sold. As a result, a parts department in a dealership is seen as stand alone business meaning it is there to earn a profit from the sale of parts to the dealership as well as to retail and wholesale customers. The management of a separate parts sales department or store is beyond the scope of this book. Consequently, the purpose of this chapter is to describe the sale of parts through a service department in a service facility.

Because parts sales at most service facilities is not the reason it is in business, the sale of parts should be perceived as a profit center and not a stand-alone business. One reason is the volume of parts sold by a facility is unpredictable. The sale of parts depends on the type of work done by the service facility as well as the kind of work that comes through the door. While a specialty service facility may have a pretty good idea of the type and number of parts sold on every job (such as an oil change or muffler specialty service facility), the volume of sales for the price that must be charged to earn enough profit at most facilities is likely to be too small and cyclical erratic to stand alone. For example, parts sales could run from 30 percent to 60 percent of a service facility's gross sales. Some services do not require parts, such as diagnostic work, while other jobs may have parts sales that make up 80 percent of the invoice (such as radiator replacement).

The primary concern for this chapter is that regardless of the work conducted by a service facility, when a part is sold, the profit on the sale must cover the costs to purchase and obtain the part. Specifically, costs begin to be incurred, even for the smallest part, as soon as the technician reports that a part is needed. This chapter will explain what these costs are and how the price of the part must be marked up to cover these costs and make a profit for investors.

Costs of Selling Parts

As soon as a parts specialist or service consultant gets a request for the price and availability of a part, he/she must call the parts store (assume a phone charge of $.25 per call). In most cases, the time on the phone to call about a part (including dial time, time to look it up at the parts store, report to customer or technician) takes at least an average of six minutes, or one-tenth of an hour of the employee's wage. The actual cost to the service facility, however, must include the amount the employer pays on Social Security, Medicare, vacation days, uniforms, sick days, medical insurance, and retirement for each employee. These expenses add about another 30 percent of the employee's pay. Consequently, if a parts specialist or service consultant is paid $10 per hour and 30 percent, or $3, is paid by the employer for Social Security, benefits, and so on, the cost per employee hour would be $13. The one-tenth of an hour taken to order the part would be $1.30 plus the $.25 phone call fee, resulting in a total cost of $1.55. This cost must be recovered in the charge for the sale of the part, assuming the repair is sold to the customer. If the sale is not made, it must be covered by the profit made from the sale of other parts and services.

After the cost of the part and its availability is given to the service consultant, customer approval must be obtained.

After customer approval, the service specialist must order the part by making another phone call and an additional cost is incurred when the second phone call (another $.25) is made. In addition to the cost of the parts and the specialist's time to order the part, time is also taken to:

- Inspect the part after it arrives.
- Record the part sale price on the invoice.
- Copy and properly file the parts delivery receipt as prescribed in the business procedures.

When all of this is considered, the part specialist can take at least another one-tenth of an hour, or another $1.30. When all goes well, the minimum cost to inquire about and later order the parts would be $2.60 ($1.30 + $1.30) of the employee's time plus 50 cents for two phone calls—$3.10 plus additional time to inspect, record, and file the information—for a total of $4.40. Not considered in this total, of course, are overhead costs (heat, light, electric, space, insurance, and so on, which can be estimated at $.50). In this case, the service facility would have a cost of $4.90 for the order of a single part. This cost must be factored into the charge made to the customer for the part.

The costs associated with ordering parts becomes greater when problems occur, including:

- Parts are not available and must be special ordered.
- Parts are wrong and must be reordered and returned.
- There is a core charge and the core's return must be arranged (reboxed, as shown in Figure 12-1, with any fluid drained out so that the residual fluid doesn't leak out).
- Defective parts must be returned and the proper paperwork must be submitted to get a part and labor cost reimbursement.

FIGURE 12-1 Cores must be reboxed and returned before a credit is issued.

- Parts cannot be delivered or delivered soon enough and a parts specialist or low-paid employee (such as a technician's helper or clean-up person) must pick up the part.

To help reduce the cost of parts, there are some cost-cutting measures that can be implemented; however, many of these measures require planning, training, and shop operational procedures to ensure that they can be conducted successfully. For example, to reduce costs, a parts specialist may:

- Look up and order multiple parts at one time. When multiple parts are ordered, the average cost per part may be less than the $4.90 discussed earlier, because it is spread over several part orders.
- Preorder parts. This will help ensure they are correct and at the service facility when the technician starts the repairs. This often will help reduce technician downtime and special emergency trips to pick up parts.
- Order parts from as few suppliers as possible to cut down on paperwork and help reduce mistakes.
- Use computers to look up, obtain price quotes, check on parts availability, and order parts online. This reduces the employee costs; however, costs must include the investment in a computer, printer, and Internet connection. Furthermore, a part specialist must become proficient in the use of a computer and know how to connect to the parts supplier's computer. (This service was used successfully at the authors' facility.)

The bottom line is that costs are created as soon as someone asks for a quote, regardless of the means used. In addition, costs may continue to be incurred once the part is obtained and after the job is finished. Therefore, cost recovery on parts sales can have surprising, unseen effects on the fiscal health of the service facility.

Gross Profit Margin and Markup

To ensure that all costs are recovered, the cost of the part to the facility must be marked up (increased). The **markup** must be enough so the service facility can meet its financial obligations plus meet the profit objectives in the strategic business plan. As discussed in the previous chapter, the labor sales area gross profit is calculated on the labor rate multiplied by the number of flat-rate hours billed to the customer, and then the technicians' wages are deducted. To calculate the gross profit on the sale of parts, the amount charged to the customer for the part (sales price) is reduced by the price the facility paid for it (purchase price). The difference between sale price and the purchase price is known as the gross profit margin, which is the same as the markup. Therefore:

Gross profit (or mark-up) = sales price − purchase price.

The costs incurred to purchase the parts (discussed earlier) and the overhead expenses (telephone, parts specialist wages, etc.) must be covered by the gross profit (or markup). In addition, the gross profit or markup also must provide a profit to meet the expectations of the owners or investors.

Assume that the purchase of a part was $10 and the sales price was $25. In this case, a gross profit of $15 was earned. The gross profit (markup) for the part would be:

Sales price	$25
Less purchase price	$10
Gross Profit (Markup)	$15

When this process is repeated for the sale of parts throughout an income statement period, the gross profit earned at the end of the period must be great enough to cover all expenses related to the parts sales area (parts specialist wages, telephone, contribution to the labor sales area for space, heat, light, electric, among other items). For example; if $2,500 was received from the sale of parts and the parts cost the service facility $1,000, there would be $1,500 in gross profit. If the related expenses (parts specialist wages, telephone, and overhead contribution) incurred by the parts sales area was $900, the net income would be $600. If they were $1,700, the net loss from the sale of parts would be $200.

In traditional accounting statements, all parts costs (cost to purchase parts for immediate resale and inventory to be kept on hand and sold) and labor costs (cost to have the technicians fix the customer's car) are

COST OF SALES	MARCH	Cummulative	Year to Date	Cummulative
Technician				
Wages	4,076.90	14.6%	8,497.75	11.8%
Overtime	39.38	0.1%	39.38	0.1%
Bonus—Tech	551.35	2.0%	710.65	1.0%
Cost of Goods Sold—Oil	479.00	1.7%	1,119.00	1.6%
Cost of Parts	4,637.30	16.6%	14,753.75	20.5%
Cost of Goods Sold—Tires	294.46	1.1%	924.85	1.3%
Cost of Sales—Other	558.45	2.0%	1,017.59	1.4%
Purchase Discounts	0.00	0.0%	0.00	0.0%
Total Cost of Sales	10,636.84	38.1%	27,062.97	37.6%

FIGURE 12-2 Cost of sales section on an income statement.

shown under the heading Cost of Sales. Figure 12-2 presents a Cost of Sales section taken from a traditional accounting statement, which was used as an example in Chapter 10.

In Figure 12-2, the cost of a part purchased for a customer's car is shown in the column Cost of Parts. If the sale to a customer comes from a specific inventory purchased for resale, such as oil, the cost of sale account is given the name of the inventory good or item sold, such as Cost of Goods—Oil. Likewise, the cost of tires sold from a tire inventory would be listed as Cost of Goods Sold—Tires. When goods are sold from a service facility's general inventory in which there are a variety of goods for sale or the same item is not sold frequently (such as nuts, bolts, light bulbs, etc.), the cost of the item sold is placed in the account Cost of Sales—Other. Separating the costs for the different items sold is important to managers for the analysis of sales. For example, if separate line item accounts did not exist, determining whether markups are adequate would not be possible. The use of a parts sales area worksheet is discussed further in the next chapter.

Parts Sales Expenses

Parts sales expenses include line items such as the parts specialist salary, parts area phone bill, computer access/networking costs, and even part of the manager's salary or a percentage of the service facility overhead when appropriate. These expenses are deducted from the gross profit on sales to calculate the net income. When expenses are greater than the gross profit, a loss is generated in the sale of parts.

Whether the $15 gross profit from this example is enough to cover the parts sales expenses and earn a profit (net income) is a difficult question. A net income or net profit on the sales of parts requires that a sufficient number of sales are made and that each sale has enough gross profit. Determining whether $15 is sufficient; however, is not impossible, if the accounting data is available to calculate the percentages, ratios, and breakeven point (explained in more detail in Chapter 14).

To illustrate how the use of percentages works, if the purchase price of a part is $10 and it is divided by the sale price of $25 ($10/$25), the cost of the part is .4 or 40 percent (.4 × 100%) of the sales price ($25 × 40% = $10). This means that if a part sold for $1, then the part would cost the service facility $.40. This also means that the gross profit percentage is 60 percent (calculated either by: the sale price, which is 100 percent, less 40 percent [cost of part] = 60 percent gross profit OR $15 gross profit/$25 sales price = 60 percent). This percentage is referred to as the gross profit margin or gross profit margin ratio. Another illustration of the same information would be as follows:

Sale Price	$25 = 100%
Part Cost	$10 = 40% ($10/$25 × 100% = 40%)
Gross Profit	$15 = 60% ($15/$25 × 100% = 60%)

If these percentages existed for the sale of all parts, this would mean that on average:

- 40 cents of every dollar received from a sale goes to the parts supplier to pay for the parts.
- 60 cents of every dollar spent on parts is gross profit and is used is to cover the expenses incurred to make the sale and for net income.

List Price and Markup

The method used to mark up parts can influence the amount made in sales. For example, some parts suppliers include a suggested sales price (called the list price) on the slip included with the delivery of the part to the facility. Some service facilities charge this suggested list price with the expectation that it is enough to cover their expenses. These suggested list prices, however, do not guarantee a gross profit that will cover the parts sales expenses of the facility. For example, the following information was recorded in a service manager's book at the end of the week.

Gross Sales:	$2,000
Cost of Goods Sold:	$1,700
Gross Profit:	$ 300

In this case, the manager expected, and really needed, $800 in gross profit to cover the parts sales expenses. He did not think he would have a problem earning his targeted amount of $800 because the previous week he sold $2,000 in parts and earned $800. The review of the manager's delivery receipts indicated that, in fact, he charged all customers the suggested list prices shown on all delivery receipts. The reason for this difference is shown in a review of a sample of delivery receipts from three different supply stores:

PART	Supplier	Suggested List Price	Purchase Price
Oil Filter	1	12.62	5.05
Thermostat	2	12.96	9.72
Air Cleaner	3	12.00	12.00

This sample showed a suggested list price that was about the same (around $12) from all suppliers but the cost (purchase price) was very different. For instance, the parts received from suppliers 2 and 3 earned less gross profit than the part from supplier 1. In a further review of the manager's record of invoices from the previous week, when a gross profit of $800 was earned, more parts came from supplier 1 than from suppliers 2 or 3. In the month when $300 was earned, fewer parts came from supplier 1 and more came from suppliers 2 and 3. In other words, if the manager wishes to use the suggested list price on the delivery receipt, he should order the parts from supplier 1; however, this may limit the type of work done at the service facility to the parts sold by supplier 1. Of course, this would lead to more complications than can be discussed here.

The answer to this problem is the manager must set a targeted gross profit margin to be made on the sale of parts. This would require that a precise method of markup be applied to the price of each part purchased. In other words, if $800 is needed on $2,000 of parts sales, a markup of 40 percent ($800 gross profit/$2,000 sale = .4 or 40 percent) on each part is needed. The manager should make his purchases from suppliers based on the lowest cost (assuming the quality is the same). The markup to give him the gross profit needed would keep the amount charged to customers at the lowest possible level. Although this illustrates the importance of using markups, it is oversimplified, as the following sections reveal.

Gross Profit Margin

When setting markups, what is initially unknown is whether a 40 percent or 50 percent or 60 percent gross profit margin ratio will pay the expenses plus earn a net income (profit). In other words, 60 percent may be right in line, or may be too low (a contributing factor to losses or low net income) or too high (a contributing to losses in sales). In some other situations, however, a facility might get more than 60 percent on some repairs because the lack of competition allows a higher-than-industry average.

Assuming the service manager has researched the industry percentage as well as what the local market will bear, and has calculated what is needed to make a desired net income (also referred to as planned profit), the parts specialist or service consultant must then be trained so that he/she can calculate markups. This is critical because to obtain the correct gross profit on each job, the correct formula must be used.

Unfortunately, many businesses use a method to calculate markups incorrectly and believe they are getting a certain gross profit margin; however, in reality, they are not. Instead, their markup method provides a much lower gross profit than needed. This, in turn, leads to unprofitable transactions or even a loss when a manager tries to discount parts to be more competitive. For example, if a $10 part is marked up to $25 and then sold at a 50 percent discount, the customer is charged $12.50. Given the costs associated with the parts specialist salary plus the other expenses discussed earlier, this would result in a loss to the service facility.

Markup Equation

The equation used to obtain the correct gross profit margin is called a retail markup. Unfortunately, although the terms markup and gross profit margin are specific, their amounts may be calculated differently by managers, bookkeepers, and accountants. Some calculations use equations that require multiplication factors, some use division factors, and some use a grossing up method (in which the cost of parts is held at 100 percent and the sale price percentage is over 100 percent). Therefore, when the term *markup* is used, one must determine the method of calculation before proceeding.

In this textbook, the mass merchandiser retail terminology is used. This means that the markup is the difference between the sales price of the part (sale price percentage held at 100 percent) compared to the price of the part. In other words, in the examples shown in this book, the percentage markup is the same as the gross profit margin percentage. To illustrate how these percentages are the same:

Sale Price	$25 = 100%	
Part Cost	$10 = 40%	the **60%** difference is
Gross Profit	$15 = **60%** ←	the retail markup

Note that the markup in this example is the same as the gross profit margin. There are, of course, transactions that cause the gross profit margin percentage to differ from the markup percentage, such as sales returns and allowances or sales discounts, but they are details in retail sales that are not relevant to this discussion.

The calculation of a markup when the cost of the part and sale price is known is illustrated in the following:

Markup percent = 1 − (part cost/sale price) × 100%.

Using the information from this example, in which the cost of the part is $10 and the sale price is $25, the markup is:

Markup percent = 1 − ($10/$25) × 100%
= 1 − .4 × 100%
= .6 × 100%
= 60%

Although the calculation of the markup for "the total parts sold in a period" is important from a management perspective, it is more important for employees (especially parts specialists and service consultants) to be able to calculate the proper markup for each part to be sold. For example, if the service facility bought a part for $10 and the owners directed management to ensure a markup of 60 percent on each part (the difference between the sales price percentage held at 100 percent and the cost of the part at 40 percent), then the following equation would be used:

Sale price = Part cost /(1 − Markup).

Therefore, if a parts specialist has a part cost of $10 that is to be marked up 60 percent or .6 (which equals the gross profit margin ratio), then that part must have a sale price of $25. This is calculated as follows:

Sale price = $10/(1 − .6) = $10/.4 = $25.

In a second example, assume the owners desire a 40 percent gross profit margin for a part that cost $35. The markup also would be 40 percent (a parts specialist would use .4 when the percentage is converted into a decimal so that it could be used in the equation). For the $35 part, the equation to work out the selling price would look like this:

Sale price = $35 /(1 − .4) = $58.33.

A mathematical proof for this example would be as follows:

Sale price = $58.33 = 100%.
Part cost = $35.00 = 60% ($58.33 × .6 or 60%).
Gross Profit = $23.33 = 40% ($58.33 × .4 or 40%).

When the calculation is done properly, the markup will mathematically equal the gross profit margin percentage. This will help to ensure that at the end of the period (provided enough sales are made) the service facility parts sale area will earn enough gross profit to cover expenses and produce a net income.

Markup Multiplier

An alternative method to calculating a markup by using division uses a multiplier. The multiplier method requires the conversion of the markup (gross profit margin), which is in a decimal form (mathematically, the decimal form comes before the conversion into a percentage) to a whole number that is multiplied by the parts cost. The equation to convert markup into a markup multiplier is:

Markup multiplier = 1/(1 − markup).

For example, assume a 40 percent markup is desired. The multiplier would be:

Markup multiplier = 1/(1 − .4) = 1/.6 = 1.67.

A markup multiplier can be found in the table of conversions shown in Figure 12-3. Note that the markup multiplier for a 40 percent markup is 1.67.

To calculate a 40 percent markup, the multiplier of 1.67 is simply multiplied by the cost of the part to get the sales price. For example, if the part cost $10, a multiplier of 1.67 would be used. The sales price would be:

Cost of part ($10) × multiplier (1.67) = sales price ($16.70).

Among the reasons that some service facilities use a multiplier instead of the standard markup equation is speed, convenience, and to avoid error. Parts specialists and service consultants must quickly

MARKUP MULTIPLIER CONVERSION

Gross Profit Percentage	Mark up Decimal	Markup Multiplier Rounded
10%	0.1	1.11
20%	0.2	1.25
30%	0.3	1.43
40%	0.4	1.67
50%	0.5	2
60%	0.6	2.5
70%	0.7	3.34
80%	0.8	5
90%	0.9	10

FIGURE 12-3 Markup multiplier conversion table.

calculate the markup for a job estimate or even for an invoice (although some computer invoice systems perform the markups as an automatic function). To make the job easier and faster, especially when a service facility wishes to use different markups for different parts, the multiplier can be extremely helpful.

Using Competitive Markups

There are occasions when markups must vary to help to increase the probability of closing a sale. For example, a service manager may find that too many muffler jobs are lost because of higher prices for parts when compared to a specialty shop. Therefore, to close more sales, the service consultants need to reduce their estimates for the repair. To lower the price, labor should not be cut because of the close margins and high expense load on the labor sales area. Therefore, many service facilities must discount their parts cost to a lower rate. This would mean if the normal markup is 60 percent, then the shop may want to drop it to 40 percent to be more competitive. To illustrate this point, assume a muffler costs $50; with a 60 percent markup, the customer would pay $125 for the part ($50 × 2.5 markup multiplier = $125). When a 40 percent markup is used, the same part would cost the customer $83.50 ($50 × 1.67 = $83.50). Hopefully, the lower 40 percent parts markup will close the sale.

There are times, however, when a lower 40 percent markup will not be enough to close a sale. This is common among more expensive items where price sensitivity is often at its greatest. For example, a $1,000 engine is less likely to sell if the gross profit margin of the service facility is 60 percent (the engine would cost $2,500 [$1,000/1 − .6 = $2,500] or [$1,000 × 2.5 markup multiplier = $2,500]). Instead, such a job might use a method in which the variable costs (costs associated with acquiring the part, such as phone calls and time of a parts specialist) are not really proportional to the cost of the part. In other words, as the cost of the part goes beyond a certain point, the variable costs do not increase as much. Therefore, these costs may be recovered through the use of a lower markup plus a markup for the fixed costs (such as facility and other constant overhead costs, which are the same for all parts sales) since a high markup may not be justified. As a result, a 30 percent margin might be used so that the engine would sell for $1,429 ($1,000/1 − .3 = $1.428 or $1.000 × 1.428 markup multiplier = $1,428), for a gross profit of $429.

When markups are reduced on more expensive parts, management must remember that the average markup percentage shown for all parts will be reduced (as well as the average percentage of the net income on gross sales). In addition, management must not drop their markups to the point where parts sales are below the breakeven point (covered in the Chapter 15). In other words, markups cost should not be cut to the point where the gross profit margin will not cover expenses.

Variable Markup

Although the high cost of expensive parts allows for easier recovery of variable costs, such as the parts specialist's time, the same is not true of lower-cost parts. For example, *activity based costing (ABC)* dictates that because of the direct cost to inquire about and order parts, such as $3.60 for direct labor and overhead, each part should be marked up a minimum of $3.60. In practice, however, this is neither possible nor practical for low-cost parts, particularly for parts bought in batches. To recover the cost of lower-cost parts and take into account the price sensitivity associated with higher-cost parts, management may choose to vary the markup based on the cost of the part. This means that lower-cost parts are marked up more and expensive parts are marked up less. To ensure that the service consultant and parts specialist choose the right markup, the manager may want to create a matrix, such as that shown in Figure 12-4.

High-Volume Part Markups

Although variable markups adjust the markup based on cost, in some situations, competition requires that certain parts be sold at a very low markup, for example, tires. Unlike expensive parts, such as engines, this lower markup is applied to increase volume. Assume that a markup of 20 percent may be obtained on each tire sale. For a tire that costs $40, the sale price would be $50 ($40/1 − .2 = $50 or $40 × 1.25 = $50). To help cover the parts sale area costs, a lower markup must be accompanied by higher volume (or substantial labor sales) to help to cover costs. Specifically, if only four tires sold in a week, the gross profit margin would not cover the costs of handling the single tire purchase. However, if the volume was greater, say 20 tires, then the volume may make up for the lower markup, because an extra $200 would be earned.

When high-volume sales of a specific part are considered a possible strategy, caution is advised. High-volume sales may cause problems that

PART COST		MARKUP	
LOW	HIGH	MULTIPLIER	PERCENT
$0.01	$2.86	4	75%
$2.87	$5.29	3.5	71%
$5.30	$9.00	3	67%
$9.01	$49.37	2.5	60%
$49.38	$110.50	2	50%
$110.51	$185.62	1.8	44%
$185.63	$533.33	1.6	38%
$533.34	$937.50	1.5	33%
$937.51	Above	1.3	23%

FIGURE 12-4 Markup table based on cost.

the service facility cannot handle. For example, an extensive (and expensive) inventory may be required or a lot of time may be needed to handle the product to the point where other job duties of the parts specialist cannot be completed. As a result, when markups drop below a certain threshold, long-term losses may occur because the sales price cannot overcome the cost associated with high-volume parts sales.

Service Facility Inventory versus Just-In-Time Delivery

Earning a profit on parts is simplest for a service facility that does not carry an inventory except for minimal supplies and small parts. Earning a profit on parts is more complicated for service facilities, such as tire dealers, that carry a large and expensive inventory. The main difference between the two is a return on the high cost to purchase the inventory must be taken into account when setting the markup. To complicate matters, however, the actual return on the investment does not come until the inventory is sold (think of inventory as money that is in a different form and when it is sold it turns back into money).

Specifically, the more often the entire inventory is sold (each time it is sold it is called a turn), the better the return on investment. This is because a little money is made on the investment each time the inventory is turned (sold, restocked, then resold again). Assume a tire dealer in a competitive marketplace has an inventory of tires that are popular with customers. Each tire costs $20 and, because of the competitive nature of the tire business, they are only marked up $2 to sell for $22 (plus the labor to install it). The return on the $20 investment per tire is earned as the inventory is turned over and over again (the tire is initially purchased, sold, restocked, then resold again). If the entire tire inventory is turned an average of 9 times per month, the return on the $20 per tire investment is $18 ($2 × 9 turns = $18) for the equivalent of an almost 90 percent return on the initial investment ($18 return on investment/$20 invested per tire = .90 × 100 = 90 percent).

However, the money made on inventory turnover cannot be considered pure profit or even a markup. For example, overhead costs must be subtracted from the return for the cost of building space occupied by the inventory, the interest paid if the money was borrowed to initially purchase the inventory, and the employee time needed to count, stock, and manage the inventory. It can be estimated that after all of the costs are considered, the actual return will be less than the 90 percent calculated and, in fact, how much of the investment return the company actually keeps is dependent on several factors, including the total volume (of tires sold) relative to the total inventory (of tires in stock). Specifically, if the tire dealer has 100 different types of tires and gets 9 turns on just 1 inventory item, it will not generate the same income as 9 turns on all 100 different types of inventory items. Other cost factors include inventory theft, inventory mistakes that cause excessive employee time to rectify (move and restock along with clerical errors), the ability of the service facility to keep in stock the type of inventory the customer wants to buy, and the ability of the service facility to adjust the inventory purchases to anticipate seasonal demands without requiring significantly more investment capital.

In general, a retail inventory can be a very complicated process to manage, particularly when a service facility must keep in stock enough inventory (known as inventory depth or how many of the same item are kept in stock) and a wide variety of inventory (known as inventory width or how many different items are in stock) to meet customer demand. When a retailer must carry a wide variety of inventory (such as high-performance tires), turns of the inventory become more complicated because some of the inventory will cost more to buy and may turn less than the average number of turns anticipated each month. In this case, the markups must be calculated so that an adequate return on investment is earned (a $40 high-performance tire that will only turn once per year cannot be marked up $2 because it will not earn enough return). The tire instead must be marked up more and, with the higher markup, it is hoped that the retailer does not overestimate the need for an item. This means there would be more of the same items in stock than can be sold before the next delivery (or the inventory never sells and sits in stock the entire year). In that case, investment capital is being wasted and would best be used elsewhere.

To help a retailer avoid this problem, inventory should be checked and marked with a sticker or chalk periodically (unless a computerized system is available). Inventory items that do not turn at least once over a period of time (typically one to three business quarters) should be liquidated so that the capital can be retrieved and used elsewhere in the business. Although overestimating the need for an inventory item causes one type of problem, underestimating the demand for an item can cause holes in the inventory, and sales can be lost because of the lack of merchandise. This means that if there is not enough of what the customer wants to buy, sales and profit from inventory turns will be lost to the competition.

Therefore, variety of inventory must be kept in stock to minimize the reasons a customer has to avoid buying the product (see Figure 12-5) and enough inventory items must be available to make the sale immediately (keeping three tires in stock when cars require four will not help make sales). This is important, because if a customer must order the product and wait for its arrival, the customer has time to get over their desire or impulse to buy the product and has time to shop the competition. This means the retailer must constantly research and estimate the correct amount and type of inventory to keep in stock. When the retailer fails to meet the customer's demand for merchandise at a given price, inventory capital will be wasted on items that will not sell and the return on the investment capital will decrease.

FIGURE 12-5 Enough inventory must be available to make the sale immediately.

Concluding Thoughts

Given the complexities of retail inventory management and sales, service facilities that deal primarily in the sale of their technical expertise should avoid extensive and expensive retail inventories. Instead, service managers should focus on training a parts specialist to help them locate, obtain, and markup for resale the required parts for customers just as they are needed by the technician to complete the job. This is referred to as *just-in-time parts delivery*.

In principle, the idea behind just-in-time parts delivery is to develop a system so that parts specialists can become proficient at getting parts for a job quickly and ahead of time when possible. This does not mean, however, that a service facility should not keep a small selection of parts that are used daily for many common jobs in stock. For example, a small inventory of most fluids such as oil, nuts/bolt/clamp assortments, oil filters, and light bulbs, among other items, should be kept in stock for the purpose of reducing the amount of time it takes to obtain these parts. A small inventory is merely for the convenience of the technicians because time is money and the business will lose more money by tying up a bay and having a technical professional wait on a $2 part, rather than to keep it in stock.

Naturally, there are exceptions to these inventory guidelines, such as rural service facilities that are far from retail parts locations or service facilities that repair cars after the part stores close. These facilities may have to keep an inventory of common parts in stock (such as brake parts, common tires, belts, hoses, and some electronic parts). Unless the service facility is in the retail sales business (such as a tire dealer) or has a dedicated parts department with personnel who understand all of the aspects of retail sales and inventory management/control (such as a new-car dealership), then a large inventory should be avoided. An owner, manager, and service consultant have enough to do ensuring that the customer is served and that transportation problems are fixed properly and promptly. They cannot have a large inventory that must be secured so that it is not stolen, and is controlled, counted, restocked, seasonally balanced, loans paid, marketed, administered, sold, and kept up-to-date. Therefore, if the delivery of parts is efficient and with a markup method that uses a multiplier conversion table shown above, a service facility should be able to move maintenance and repair jobs through their shops without any extended time delays and without inventory complications.

Review Questions

1. What costs are incurred by a service facility when selling parts?
2. What are the different methods used to mark up for parts? Give an example of each.
3. What is meant by just-in-time parts delivery and what is its advantage?
4. Why is the gross profit margin for parts sales and the markup the same? Why would they not be the same?

CHAPTER 13

FINANCIAL MANAGEMENT OF A PARTS PROFIT CENTER

LEARNING OBJECTIVES

Upon reading this chapter, students should be able to:

- *Define profit center.*
- *Discuss the responsibility of a service manager to the labor and parts sales area profit centers.*
- *Describe the format and purpose of the parts sale management worksheet.*
- *Explain the target metrics for the line item accounts on the parts sales management worksheet.*
- *Discuss the analyses to be conducted with the information presented in the parts sales management worksheet.*

Introduction

We have assumed up to now that the service manager is responsible solely for the labor maintenance, diagnosis, and repair operations, whereas parts sales are handled in another department under another manager. In a number of facilities, however, the service manager is responsible for the sale of parts, which is the focal point of this chapter. When the parts sales area is under the service manager, essentially the manager becomes responsible for the financial oversight of two areas: labor *and* parts sales. This chapter presents and describes the sources of income, cost of sales, expenses, and related concerns in the parts sales area. In addition, the chapter explains the composition of a parts sales worksheet and how it is used to monitor the parts sales operations.

Another purpose of this chapter is to point out that when a service manager has two areas or operations to supervise (labor and parts sales), worksheets should be prepared for each area. This requires that the incomes, costs, and expenses of the facility be divided between the two areas. When two or more areas are under one manager, they are referred to as *profit centers*. As a result, the chapter defines a profit center and explains why and how income, costs, and expenses are assigned to profit centers and their worksheets.

What Is a Profit Center?

Service facilities, for example, a large dealership, are likely to have separate departments, such as a sales department, a service department, and a parts department. The reason they have departments is so that the owners or stockholders and upper managers can keep an eye on each department's financial performance. This subdivision is also necessary when a company gets too large and it becomes too complicated for one person to manage. In many cases, a service facility is not large enough to justify having separate departments with a manager in each department. However, most service facilities are large enough that the financial performances of services sales and parts sales must be separated otherwise they become blurred. In other words, more managers are not needed, but when all of the incomes, costs, and expenses are lumped together, the manager in charge cannot tell where the facility is making and losing money. Specifically, the manager must know whether the facility is making a profit on the sale of labor and the sale of parts.

Therefore, instead of departments or divisions, a service facility can set up profit centers for the labor sales and parts sales areas (see Figure 13-1). The centers come under the same service manager, who has control over the income and expenses of each center. In other words, a profit center is a specific business activity. This means that at most service facilities (except dealerships and some specialty shops) a manager is responsible for more than one profit center. There usually is a labor sales profit center and a parts sales profit center with one manager responsible for both centers. Owners expect that the manager will be responsible for each area and will ensure a profit is earned at each center.

As will be discussed in Chapters 14 and 15, profit centers may be created for other areas. For example, a facility may devote an entire bay to alignments. From a sales standpoint, it may seem logical that the amount charged for an alignment is based on the stated labor rate and when the cost of the technicians labor is deducted, a gross profit will result. This may lead a manager to believe the alignment area is profitable. However, this may not be true and cannot be determined unless the alignment sales are separated out (as should be done on the final invoice). When the manager subtracts from all of the alignment sales for the month, the cost of the labor needed to operate the machine (this is the gross profit margin) as well as the recovery of the cost of the expensive alignment machine and special alignment rack, the maintenance of the machine, and the overhead (space, heat, light, electric, management oversight, etc.) it uses, alignments may not be profitable. In other words, setting up a profit center for the alignment service with all costs accounted for on a worksheet (similar to the one in Chapter 11) is needed to determine whether alignments are making a profit. The details for setting up multiple centers in a service facility are discussed in the following two chapters along with the course of action to be taken when a center is found to not be profitable reviewed.

Creation of Profit Center Worksheets

As explained in the earlier chapters on accounting, entries are made into the accounting records of a facility for all business transactions. There is only one set of books and only one set of statements (income

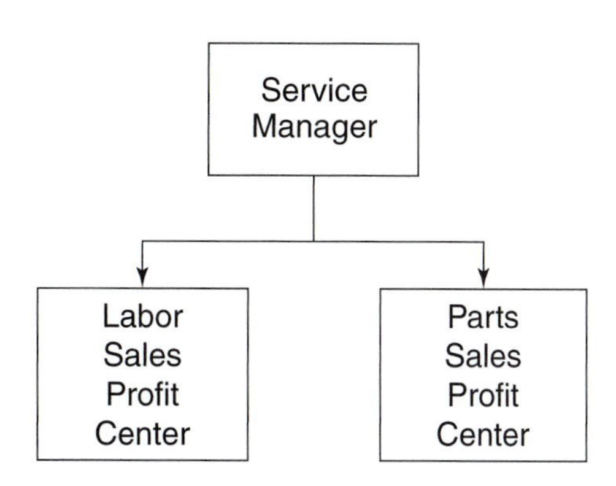

FIGURE 13-1 Profit centers in a service department/unit.

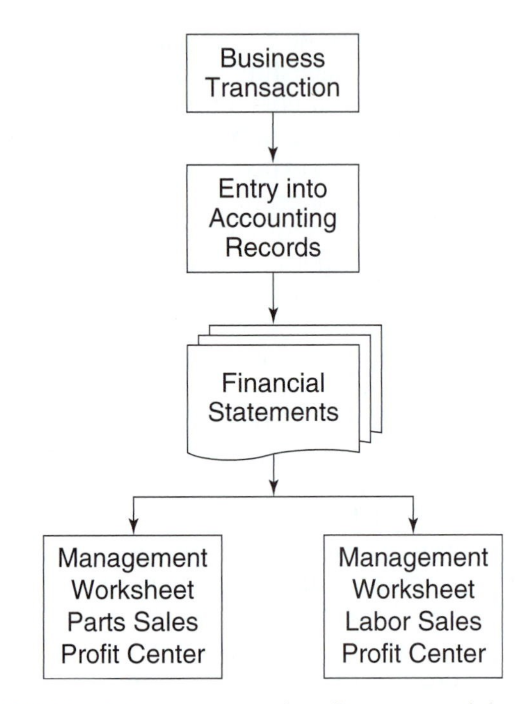

FIGURE 13-2 Generation of profit center worksheets.

statement, balance sheet, and cash flow statement) for a company. Figure 13-2 reviews the process from transaction to account entry to the traditional financial statements (income statement, balance sheet, cash flow statement). These traditional financial statements are used to present the financial performance of the business to the owners, federal, state, and local government, and banks, among other external users. The income and expenses found in the income statement are then copied into management worksheets of profit centers. Therefore, a profit center worksheet is not an official document but, rather, a working document used for internal management purposes by those who must make decisions about the direction of the company and must be able to accurately monitor each center's operations.

Allocating Income, Costs, and Expenses to Centers

When two or more profit centers are established, the incomes, costs, and expenses must be properly allocated to each center. For example, a labor sales area worksheet presents all labor sales, cost of labor sales (technician wages) and expenses of the service department (see Chapter 11). When parts sales are added, a profit center for parts sales must be created, which requires another worksheet be prepared. Therefore, the labor sales area becomes a profit center with its own worksheet and the parts sales area also becomes a profit center with its own worksheet (discussed later in this chapter).

Allocation of Income to Profit Centers

Income assigned to each profit center is the money earned from the sale of services or goods by that profit center. The amounts of the income are obtained from customer invoices. Specifically, each customer invoice should indicate how much the person paid for labor and how much was paid for parts. The amount paid for each is then entered into the accounting records and their total appears on the income statement (see Chapter 10 income statement for Renrag, Inc.). The income of each profit center is taken from accounting records and transferred to the appropriate profit center worksheets. For example, the total amount shown for labor sales on the Renrag, Inc. income statement would appear in the worksheet of the labor sales profit center, whereas the totals for sales—parts and accessories, sales—oil, sales—other, and sales—tires would appear on the profit center worksheet for parts sales. Figure 13-3 is an example of an area set aside for a sales—tire profit center which could even become a suboperation, as shown in Figure 13-4.

If the managers of Renrag, Inc. felt that they needed to break down the labor sales profit center further, they may decide to make another profit center (and another worksheet) such as a lube-oil profit center. In

FIGURE 13-3 The sales profit center.

FIGURE 13-4 Sales—tire suboperation.

this case, the labor for the lube-oil operation as well as the oil and filters sold to customers would both appear in that profit center. In other words, the lube-oil profit center worksheet could have both labor and parts associated with the service in one worksheet. Naturally, when oil is sold for purposes that are not associated with the profit center, such as an engine repair, the sale would be allocated to the parts sales profit center. This will allow an accurate account of the income from the sale of oil by the lube-oil center so that gross and net profits can be calculated properly.

Cost of Sales to a Profit Center

As in the allocation of income, the cost to pay for the service (labor) or purchase the goods (parts, oil, etc.) must be assigned to the appropriate worksheet. In some cases, an expense line item may need to be split proportionally and entered in different worksheets. For instance, the sale of the oil for the engine repair in the example discussed in the previous section was assigned to the parts profit center and not the lube-oil center. Likewise, the cost of the oil for the engine repair also should be charged to the parts sales worksheet. In other words, the accuracy of the charges is important and must be overseen and supervised by the manager. Otherwise, the worksheet will have limited value to the manager to determine the profitability of the profit center because the information presented in the worksheet is not assigned accurately and, therefore, not valid.

Allocation of Expenses to Profit Centers

When expenses are shared by profit centers, such as overhead for heat, light, rent, and so on, they should be proportionately charged to each profit center. For example, each profit center would be charged for electricity. Although it would be ideal to directly measure the amount of electricity used by each center via separate electric meters, it is not practical. Therefore, the manager must determine how the amount should be split. One method is to base the charge on the percentage of space each center occupies. For example, if the labor center occupies 90 percent of the space, it would be charged 90 percent of the electric bill, whereas the parts profit center would pay 10 percent. At the same time, because the parts center would not use any water, except for the water in a drinking fountain, the entire water bill may be charged to the labor profit center.

The wages of the service manager also could be shared. A portion of the wages paid to the service manager would be for supervision and assistance (a service consultant's wages also could be split among different profit centers depending on the duties assigned). Although the amount charged may not much, it is important to recognize because it signifies to the manager that a portion of his/her pay is earned from the sales of the parts profit center.

Another method to assign costs is to base it on time worked in each profit center. For instance, a wage payroll person may be assigned to one profit center for various amounts of time or may perform overlapping duties such as a drive customers to work then pick up parts on the way back to the service facility. The hours worked for each profit center should be recorded, assigned to the proper profit center, and charged on the proper profit center worksheet.

In application, as profit centers are added, the charges also must be assigned appropriately. In some cases, the charges may be based on the amount of space used and the amount of time the space is occupied. For example, assume a service facility has two profit centers: general repair operations and a tune and lube operation (both in the same building sharing space). The general repair shop occupies 60 percent of the space and the tune and lube center occupies 40 percent of the space. A manager must determine how to allocate a property tax bill (this is a constant charge that does not increase with use) for $100. When only space is considered, the general auto repair operation would pay $60 and the tune and lube operation would pay $40. However, for a bill that is not a constant charge, such the use of heating oil, space alone may not be enough and operating hours may need to be considered. Had the tune and lube operation been open longer hours (evenings and weekends), it would use more heating oil. Therefore, the charge for heating oil may be based on both hours of occupancy and use of space. An adjustment would be made for the amount of time the tune and lube operation occupied the building. For example, because it is open 20 percent more hours per week than the general repair operation, an additional charge of 8 percent

(20% × 40%) of the total heating oil bill would be paid by the tune and lube operation with 8 percent less paid by the general repair operation. For a $100 heating oil bill, the tune and lube operation would pay $48 ($100 × (.4 + .08) = $48) and the general repair operation would pay $52 ($100 × (.6 − .08) = $52). Although there could be other methods to split this expense that might be more fair, the point is that expenses must be accurately assigned to each profit center.

Purpose of a Parts Profit Center

The purpose for treating the parts sales area as a profit center is simply to determine if it is making a profit. Specifically, the parts profit center must be able to cover a wide variety of daily operating expenses and must earn a modest profit (hopefully enough to help to cover business loans and provide the owners with an acceptable return on their investment so they can continue to obtain the capital to grow the business). It is the responsibility of the service manager to ensure the business earns a consistent profit in *both* profit centers. The profit of one profit center cannot cover a loss in the other. It can be worse when a service manager does not know when one center is making a profit and the other is suffering a loss. Although a service manager may feel that, as discussed in the last chapter, the markup of each of part sale is enough to ensure a profit, in reality this may not be so. There must be enough total income from the sale of parts to cover the cost of the parts, the expenses associated with the sale of the parts, and earn enough profit.

Responsibilities of the Parts Specialist

The parts specialist is a support staff member who also may serve as the service consultant in smaller service facilities. In general, the parts specialist's responsibilities, some of which may be performed by the service manager, are to:

- Obtain quality parts at a fair price and just in time to maximize the efficiency of the technician.
- Keep track of small inventory items such as nuts/bolts/clamps, oil, filters, and other fluids.
- Ensure that each part, fluid, and small inventory items used by the technician is relayed to the service consultant so it can be charged to the customer's invoice.

- Acquire and distribute to the technician needed supplies, such as general shop fluids (e.g., brake cleaning), shop rags, floor mats, seat covers, hang tags (on mirror), and cleaning supplies among other items.
- Administer environmental matters related to the operation of the service facility, such as waste-oil removal and cleaning solvent among other items.
- Ensure tire disposal fees are adequate to cover the cost of tire removal and proper disposal in accordance with all local, state, and federal laws.
- Arrange and ensure that any repairs and services performed by outside vendors (such as radiator repairs) are marked up properly and provided to the service consultant for the invoice the customer.
- Arrange to have scrap metal from repair operations at the service facility sold or recycled.
- Make sure parts suppliers' delivery receipts and returns are accurate (returns and cores are credited properly to the account) and filed properly in order for the bookkeeper to process checks in a timely manner for a 2 percent/10, net 30 discount.
- Perform all paperwork for the collection of labor (when applicable) and parts payments from suppliers or manufacturers for defective parts sold to the facility.
- Collect and record all other income earned such as the soda, snack machines, and shipping charges.

The typical method of payment for a parts specialist is hourly wage. Some service facilities, however, may provide a bonus based on the sale of parts or, preferably, on the profit made from parts sales. The payment of a bonus based on profits is to ensure the part specialist focuses on the reduction of expenses and not just an increase in sales. This would mean that it is in the parts specialist's personal interest to make sure returns and cores are credited properly and returned in a timely fashion, delivery receipts are correct, and parts are charged out on the invoice for an amount that is marked up correctly.

The Parts Sales Area of Worksheet

In order for a service manager to track the performance of the parts sale profit center, a parts area worksheet is used (see Figure 13-5). This worksheet is divided into four sections with subsections as shown in the following outline.

1. Parts Sales Area Income
   - Income from Customers
   - Income from Other Sources
2. Cost of Parts

3. Expenses
   - Support Staff
   - Employee Benefits
   - Operations
   - Overhead
4. Net Income

In the parts sales worksheet (see Figure 13-5), the amount of the income or expenses are shown in the first column with their percentages presented in the second column. For example, in the income section, the gross parts sales was 90.89 percent of total sales; therefore, the income from other sales was nearly 10 percent, or over $2,000. The cost of parts amounted to 39.82 percent of total sales, which is acceptable because it is less than the target metric set at 40 percent, and the net income amounted to 34.43 percent of total sales, which is acceptable because it is greater than the target metric. If the net profit should dip below 30 percent, the service manager can easily examine the target metric to determine which income or expense examine the target metric to categories were out of line or if it were an income problem.

Parts Sales Area Income Section of Worksheet

As shown in Figure 13-5, the income from parts sales comes from customers and a variety of other sources. Although most income is from the resale of parts to customers, the other money received from customers and other sources cannot be ignored. In some cases, the income received from a source may be quite a bit greater than shown in the example. For example, some facilities may have a large volume of business generated from sublet repair sales. This suggests they consider expanding into this service, such as alignment services and target small auto repair shops that do not have a machine. In other cases, a facility may not be taking advantage of some of the sources of income, such as money that can be received from scrap metal (used lead wheel weights), shipping fees that should be added to a customer's invoice, and fees for tire disposal. While these may not be large amounts of income, the failure to collect them is bad business and a facility can be nickel and dimed to death. In other words, over time these incomes add up to significant amounts.

Gross Parts Sales

As discussed in the last chapter, markup is one of the greatest concerns for a service manager. When the service manager establishes a markup method, it must be monitored. A parts specialist must be able to make the correct calculation and always apply it. For example, a customer who worked for a professional baseball team wanted to trade free game tickets

PARTS SALES AREA INCOME INCOME FROM THE CUSTOMER	Receipts	Percentage	Target Metric
Gross Parts Sales	$18,765	90.89%	
Shop Supplies and Environmental Administration Fees	$777	3.76%	
Shipping Fee (charged to customer as a separate line item)	$125	0.61%	
Tire Disposal Sale	$95	0.46%	
Sublet Repair Sale	$300	1.45%	
Total Customer Income	$20,062	97.18%	
INCOME FROM OTHER SOURCES			
Scrap Metal	$33	0.16%	
Part Discounts and Promotions	$160	0.78%	
Part Supplier Payment (for labor to install defective part)	$175	0.85%	
Other Income	$215	1.04%	
Total Other Income	$583	2.82%	
Total Income	$20,645	100.00%	100%
COST OF PARTS			
Part Purchases (without 2%/10 net 30 discounts)	$7,478	36.22%	
Shop Supplies Purchased (separate Invoices)	$430	2.08%	
Tire Disposal Cost	$67	0.32%	
Payment to Labor Sales Area for Defective Parts	$0	0.00%	
Cost of Shipping (charged to customer as Shipping Fee)	$55	0.27%	
Cost of Sublet Repairs	$190	0.92%	
Total Cost of Parts	$8,220	39.82%	<40%
Gross Profit	$12,425	60.18%	>60%
EXPENSES			
Support Staff			
Parts Specialist Wage	$1,720	8.33%	Hourly wage
Service Manager Wage (partial)	$1,200	5.81%	optional
Commissions/bonus	$300	1.45%	optional
		7.7%	of staff
Social Security / Medicare Tax	$248	1.20%	wages
Employee Benefits			
Uniforms	$50	0.24%	$50/employee/mo
Retirement	$170	0.82%	optional
Medical Insurance	$200	0.97%	optional
Development Programs	$0	0.00%	optional
Operations			
Equipment Leases			
1	$0	0.00%	optional
2	$0	0.00%	optional
Information Technology	$75	0.36%	1–2% of gross sales

FIGURE 13-5 Parts sales profit center management worksheet.

PARTS SALES AREA INCOME INCOME FROM THE CUSTOMER	Receipts	Percentage	Target Metric
Other Expenses	$55	0.27%	1-2% of gross sales
Office Supplies	$22	0.11%	1-2% of gross sales
Telephone Expense	$57	0.28%	<2%
Travel Expenses	$0	0.00%	<2%
Waste Removal Fees	$275	1.33%	7%
		0.00%	
Overhead			
Rent Contribution	$400	1.94%	0.4 to 0.6/sq ft/mo
Property Tax Contribution	$50	0.24%	.05 to .08/sqft/mo
Heat / Light / Electric Contribution	$120	0.58%	.12 to .2/ sqft/mo
Insurance Contribution	$200	0.97%	$200/employee/mo
Payment to Labor Sales Area for Defective Parts	$175	0.85%	optional
Total Expenses	$5,317	25.75%	<30%
NET INCOME	$7,108	34.43%	>30%

FIGURE 13-5 *continued*

for a reduction in the cost of the parts put on his car. The owners (authors) said no; however, they learned later that the parts specialist took the offer (he had to pay the facility for the reduced cost on parts). In another case, the owner of an automobile was given permission to bring his own parts to the facility without the owners' (authors') knowledge. The authors just happened to stop in when the customer was returning from his third trip to the auto supply store to get the correct parts. The loss on the profit on the parts was not as great as the loss of bay time from the car sitting on the lift waiting for the customer to return with correct parts!

Another reasonable target metric is for parts sales to average around 40 to 60 percent of the labor sales at most independent service facilities. In other words, if the gross labor sales of an automotive repair business was $20,000, gross parts sales to the customer should average $8,000 to approximately $12,000. If the gross profit on parts is expected to be 60 percent, then the amount of gross profit should range from $4,800 to $7,200.

Income from Shop Supplies

In addition to the parts sales, income also must also cover the cost for:

- Shop supplies
- Shop rags
- Floor mats
- Seat covers
- Hang tags (on mirror)
- Computer access fees to the parts stores
- Cleaning supplies

- Administration of environmental matters related to the business (such as contracting for and maintaining files that track waste removal and solvent tank maintenance manifests)

To track and itemize each charge for every repair job is not really possible. Therefore, some service facilities add a charge for these items by using an account called Shop Supplies/Environmental Administration Fees. The charge for this fee may be between 1 and 4 percent of the total parts and labor sales found on the invoice. For example, if the customer bill is $100, the customer would be charged anywhere from $1 to $4 extra to cover these items. Other service facilities may add a fixed amount of $1 to $2 to the labor rate to cover these costs. The receipt is then transferred as fee income to the parts sales area. At some service facilities, the parts sale area will try to recover the supply costs by charging the customer for each supply item, such as a can of brake fluid even if a full can was not used. Whatever method is used, the point is that the parts sales area must recover these costs but it should be done fairly; otherwise, customers may be lost.

Shipping Charges

Purchasing parts and paying the cost to have them shipped to the service facility is quite common. In these cases, the shipping cost is considered a cost of sales and recorded in the cost of parts section as Cost of Shipping. To charge the customer for the shipping, the facility may handle it in one of three ways.

The shipping cost for a part or parts for a single customer can be added to the cost of the part sold to the customer. In other words, the cost of the part shown on the invoice is simply increased to cover the shipping and the total amount (part cost and shipping charge) is marked up; however, a record of the money received for shipping will not appear in the accounting records of the management worksheet.

The shipping charge to the customer can be shown as a separate item on the customer's invoice. The amount charged may include a markup and the total charge recorded on the customer's invoice as shipping charges. The benefits to this method are that it offers the opportunity to use a different markup for shipping, it reduces the cost shown for the part on the invoice, and, if the part is returned, the shipping will not be included in the reimbursement (only the charge for the part). Of course, in some facilities the policy is to not offer refunds on special-ordered parts.

When many items (such as small hardware, nuts, and bolts) are shipped into the facility at the same time and it is difficult to attach a charge to a particular item, then it is best to expense the shipping charge under the line item shipping expense.

The second method described above is used in the parts sale area worksheet (see Figure 13-2). Specifically, the cost of the shipping is recorded in the cost of parts section while the shipping fee charged to the customer (after the proper amount of markup has been added) is recorded in the Income Section.

Fees for Tire Disposal

Disposing of tires is a problem. As a result, income should be collected from customers for tire disposal. These disposal fees may vary from $1 to $5 per tire depending on the size of the tire, location of the service facility, and charges of the legally authorized collector. Tire disposal is a problem because it can be difficult to find a legal hauler. When a hauler is not licensed and insured according to state law, the service facility is liable for the cleanup if they are not disposed of properly. Because a tire hauler may change prices from one visit to another, coming up with a standard fee to charge customers is a problem.

Sublet Sales

Given that some equipment, such as a tow truck, is not owned by all service facilities, a service facility may contract with another facility to conduct the work. The parts specialist is usually the person who makes the arrangements for these services. For example, if a customer must have a car towed to the facility for repair, the service consultant or parts specialist may contact a towing service to pick up the vehicle. The towing service usually invoices the facility on a monthly basis and the charge is entered into the cost of parts section as Cost of Sublet Repairs. The facility then adds a fee to the customer's invoice for making the arrangements and for the towing service. The amount of the markup is usually a minimal fee that covers the charge plus expenses and not a markup.

For other sublet repairs, a markup may be included. For example, if an employee must drop off and later return to pickup a radiator sublet repair across town, the cost of the service is greater than picking up the phone and calling a tow truck. To recover the greater costs of sublet repairs and earn a profit on the labor and supplies for taking the part off, checking it, and installing it may require a labor charge or larger markup (say 70 percent, but the markup should not make the cost of the repair greater than the cost of a new part, such as a radiator, radio, or transmission). Another approach is to use a normal parts markup on all sublet repairs, say 40 percent, except when excessive employee time (labor and delivery) and use of the company vehicle is involved.

Scrap Metal

Gross parts sales income also can be bolstered by money collected from sources other than the customer, such as selling scrap metal obtained from the repair process. Although the monthly income from scrap metal is minimal, it often can add up to several hundred dollars over a year, provided it does not require a large amount of employee time or excess space for storage. Figure 13-6 shows a special dumpster for scrap metal.

FIGURE 13-6 Scrap metal dumpster.

Income from Manufacturer/Supplier Discounts and Promotions

Another income item is actually a savings from discounts (2 percent/10, net 30) and manufacturer promotions on parts. Although some service facilities might treat this as a subtraction from the Cost of Parts, in reality it must be seen and treated as income. The reason is some promotions offer a money-back program and a check is sent directly to the service facility for selling a specific brand, such as tires or brakes.

With respect to the parts supplier's 2 percent/10, net 30 discount (2 percent discount when the previous month's invoice is paid by the tenth of the month; otherwise the full net amount is due by the thirtieth of the month), the savings does not show up on the parts supplier's invoice. It must be subtracted by the bookkeeper and then the discount is entered into the accounting records. Therefore, for record-keeping purposes, the discount is recorded as income rather than subtracted from the cost of parts. To earn this discount, the parts specialist should quickly and accurately check over the supplier invoices each month so that a check is generated by the business and delivered to the parts supplier within the discount period (within the first 10 days of the month). For accuracy and analysis purposes, the amount save should be treated as income for the parts sales area and not subtracted from the parts purchases.

Income—Replacement of Defective Parts

A parts supplier invoice will not only indicate an end of the month discount, it will show any parts and labor allowances or credits for the replacement of defective parts supplied by their store. To ensure that the service facility obtains credit for its claims, the parts specialist must follow the parts supplier's procedures in verifying the problem, returning the defective part, and making follow-up calls to ensure all necessary paperwork is properly submitted. Then the parts specialist must inquire about the reimbursement periodically until it appears on the part supplier's invoice. Typically, "big-name" brand parts suppliers provide a labor reimbursement for the installation of defective parts and, when the claim is paid, the labor should be recorded as income or transferred internally to the labor sales area worksheet as income.

Other Income

Although other incomes could be treated as a separate profit center, the amount is usually not enough money to justify the trouble of setting up the accounts. Therefore, other income received from various sources for a number of reasons is accounted for in the parts sales worksheet. For instance, a service facility may have vending and snack machine income and people will stop into a facility and buy a quart of oil, some antifreeze, and so on. Other income sources such as end-of-the-year gifts from suppliers as well as trades for services between businesses (there will likely be a labor sales income to trades as well as a parts sales income).

Cost of Parts Section of Worksheet

The section on the Cost of Parts involves those expenditures that directly related to the sale of parts. Unlike expenses, these costs should show some relationship to the income accounts. For instance, how does the cost for tire disposal compare to tire disposal sales? How does sublet sales compare to the cost for sublet repairs? In other words, although the percentages of the costs to the total income are important, the worksheet is also set up to permit direct comparisons that may be even more important to the service manager.

Parts Purchases

The cost of the parts purchased each month is obtained from the invoices received from the parts suppliers. The danger in making a comparison of this amount to gross parts sales up through the last day of the month; however, the date of the supplier's invoice is usually not the last day of the month. As a result, the amount of the parts purchases must be adjusted for the management worksheet and the amount may still be off; however, the worksheet is a working document and not a formal statement that guarantees accuracy. For instance, because the last delivery on the invoice may be the twenty-fifth, the amounts of the delivery receipts from the twenty-fifth to the last day of the month must be added to the invoice total plus the amounts on the invoice for the last days of the previous month (say the twenty-fourth to the last day of the month) must be subtracted from the total. In some cases, a job that spanned the end of the month into the next month may have to be added or deleted from the total parts purchased. In addition, the amount of the cost of parts should not include the labor/parts payments received from suppliers for defective part credit and the 2 percent/10 net 30 discounts should not be subtracted from the total because it is recorded as parts sales income. Also, the parts purchased amount should include for inventory items from the shop's inventory used in repairs and not charges for items on the parts suppliers' invoice that went into inventory, such as air and oil filters, and were not used during the month.

Shop Supplies Purchased

The Cost of Parts section also includes an account Shop Supplies Purchased. These shop supplies may be included in the suppliers' invoices (noted earlier) or a special invoice when purchased from a vendor who only sells supplies. For example, a vendor may come to a facility and maintain a supply inventory of oil, small parts, and air filters. Shop supplies also may include cleaners (brake cleaner, hand cleaner, floor soap, etc.) as well as rags, seat covers, floor mats, and other miscellaneous expendable supplies needed to complete a job (see Figure 13-7).

The reason for maintaining a record of the supply inventory and a separate account on the worksheet when they were used in a sale,

FIGURE 13-7 Shop supplies, like floor mats, may be included in the suppliers' invoice.

however, is the tendency for them to be improperly handled and wasted. For instance, a can of brake fluid often may be half-full sitting on a shelf or toolbox. When brake fluid is needed the technician will open a new can for a repair instead of using the half-full can. The brake inventory may disappear and the records may indicate that only half of them were used in sales. Therefore, ideally a separate set of inventory records and the use of an account when inventory is sold provides a means to monitor, control, and analyze the use of these sometimes costly items; however, the method can be too time consuming.

Because the separation of parts supplies from specific parts purchased for a repair (such as a set of brake shoes) is difficult to separate when delivery receipts are compared to invoices for control purposes, the parts specialist should request parts suppliers to prepare separate delivery receipts showing only shop supplies. The other alternative is to periodically order shop supplies on a separate order. In these cases, then, the inventory and shop supplies should be checked and replenished each week, such as the end of the week for delivery on Monday morning.

Tire Disposal Cost

As explained earlier, used tires must be hauled away from the premises and disposed of in a legal manner. This is not a job for which the vendor charging the lowest fee should necessarily be given. If it is not done legally, the service facility will be liable for government cleanup charges plus any fines. Although this fee for removal is suggested as a cost of parts for analysis purposes, it also could be considered an expense line item because of the legal issues regarding the disposal of the tires. The classification is a decision to be made by the owners; however, because it was a direct charge to the customers (Tire Disposal Sales) the authors entered it under parts purchased.

Shipping of Parts for Customers

The next item in the Cost of Parts section is the Cost of Shipping charged to customers for special-ordered parts (discussed earlier under Parts Sales Area Income). Again, shipping costs are typically paid by the facility when a part is special ordered. The recommendation, however, is for the customer to prepay the shipping costs along with a deposit on the purchase of the part. For example, if a customer wishes to purchase an engine for his/her automobile, the person should pay a nonrefundable fee that would include the shipping of the part to the facility and a deposit that can cover the shipping of the part back to the manufacturer in the event the customer changes his/her mind.

Sublet Repair Costs

These costs coincide with the sublet repair sales discussed earlier and represent the cost of work done by another facility. A markup of the repair should be applied to cover the costs of the duties performed by the parts specialist, such as:

- Arranging for the vendor to perform the repair
- Arranging drop-off/pick-up of the vehicle or components
- Shipping and return shipment of the component to be repaired
- Arranging (issuing purchase order numbers) for vendor payment
- Collecting outside vendor receipts for billing purposes
- Setting up a warranty for the repair for the facility

Parts Sales Area Expense Section

In the process of acquiring parts to finish the customer's job, an automotive service facility generates business expenses, some of which are shown in Figure 13-2. Although some expenses are directly related to running the parts area, such as the parts specialist's wages and benefits, equipment leases, information technology, office expenses, telephone, and waste removal, other expenses are shared with the labor sales profit center or other profit centers. Shared expenses are considered to be overhead expenses, such as rent or the building mortgage payment, heat/light/electricity, and property taxes. In other words, overhead is necessary to conduct business but the expense line items cannot be directly connected to one profit center or activity.

Parts Specialist Wages, Commission/Bonus, and Benefits

As explained earlier, the parts sales profit center would employ a parts specialist and also be charged for a portion of the manager's salary as shown in the worksheet (see Figure 13-3). In addition to these salary charges, the employer must make the standard Social Security contributions on all wages and the profit center should be charged for all benefits. Because the specialist would come into contact with customers, he/she should be provided with a uniform that is similar to or same as the service consultant as well as development opportunities.

Parts Office Supplies

This item includes office supplies such as pens, paper, photocopier paper and cartridges, computer printer cartridges and special paper (if needed), purchase order pads, file cabinet folders, and so on. Often, the parts specialist maintains the files for the service facility including customer files and folders on shop equipment; however, these expenditures should be charged to the labor sales profit center. In addition, the parts specialist may take care of and the parts sales center pay a portion for the photocopy machine (repairs and servicing), parts catalogs, computers and software that interface with the parts suppliers, and the server.

Telephone Expense

Telephone expense charges for the parts sale area consist of the phone lines to the service specialist work station plus a sole connection from the parts specialist's computer used to contact parts suppliers. The phone line for the parts specialist must be independent of all other lines (no other phone at the facility should roll over to the parts phone line). Because it would be usual for the parts specialist to make long-distance calls, a log of all calls should be kept and checked against the charges for long-distance phone calls by the phone company. These calls also need to be monitored because they can unintentionally get out of hand quickly, especially if customers or employees can gain unsupervised access to the phones.

Travel Expenses

Although most parts suppliers deliver parts, there are occasions when travel to pick up parts and other materials is necessary. There are two methods used to make travel reimbursements. One method is to reimburse employees on the actual costs of the travel (gas receipts, maintenance costs for the automobile, etc.) and the other is to use the number of miles traveled.

In 2005, the federal government permitted $.42 per mile be paid to employees for the use of a car. The reimbursement is the recorded as travel expense for the facility. For example, if the parts specialist or any other employee used his/her automobile and traveled 100 miles to pick up parts from various vendors, the reimbursement for the cost of the travel would be $42 (100 miles × .42 per mile = $42) plus any tolls and parking fees paid. The facilities insurance carrier should be consulted about the coverage on employee automobiles when used for the business.

When a business trip requires overnight lodging and meals, these expenses are usually covered by the employer but for a maximum amount, such as $100 for a room, $5.00 for breakfast, $7.50 for lunch, and $12.00 for dinner. For long-distance travel, the expense reimbursements also may include airline tickets, taxi service, and rental cars. In some cases, a facility may wish to make arrangements for lodging as opposed to reimbursing the employee in order to gain a business discount and maintain control over the costs.

Other Expenses

The other expenses category is for expenditures that do not fit into one of the expense line item accounts. For example, the parts sale profit center may have to pay fees to conduct business with a certain vendor (such as a membership to purchase supplies from Sam's Club) or to be able to buy special software to update their computer database on the latest manufactured automobiles.

Waste Removal and Environmental Administration

This item does not include trash removal but, rather, the removal of waste oil, used antifreeze, and cleaning solvents by approved companies. In other words, the line item expense is a cost charged against the gross profit made on the sale of these services. This requires the maintenance of records and manifests of approved haulers (such as Safety Clean) as well as the submission of reports to government agencies as required by state and local laws. A duty of a parts specialist is to order these services and keep on file the necessary paperwork about the company's chemicals (such as the Material Safety Data Sheets) in order to verify that the facility is in compliance with all local, state, and federal laws. In some cases, consulting companies can be paid to assist in these endeavors and, when this is the case, their fees would be charged to this line item expense account.

Overhead

Because a parts sales profit center and the labor sales profit center share space and some general expenses incurred to conduct business, each must pay its share. Known as overhead expenses, a portion of the expenses, such as rent and heat/light/electric, should be partially covered by the gross profit from the sale of parts and from the sale of labor. These charges can be shared by each profit center. The method, percentage, or amount charged to each profit center is usually determined by the manager.

Net Income

The net income is determined when the cost of parts and the expenses are subtracted from the sales income (see Figure 13-2). After the net profit is calculated, the percentage of each category relative to the total income of the parts sales area is calculated by dividing each section total into the total income. Then the percentages for each line item and section of the worksheet must be analyzed and compared to the target metrics. When a percentage is outside of the targeted range, the analysis worksheet may conduct additional calculations to help uncover possible problems as discussed in the following two chapters.

Final Thoughts

The form to be used for a profit center worksheet may be included with accounting software or can be prepared through the use of spreadsheet program software, such as Microsoft Excel. The worksheet should calculate the percentages shown in Figure 13-3 and, in some accounting software programs, it may make comparisons between the performances for the current month and year to date to those of the previous year.

One issue to be recognized when a facility has two profit centers regards the labor reimbursements from the parts sales profit center to the labor sales profit center. This may be a controversial issue because the labor reimbursements, in terms of the amount charged per hour or the time allocated for a repair, may be lower than those charged by the facility. This results in a lower gross profit and a possible loss on the job for the labor profit center. In some situations, an inadequate reimbursement may be a result of errors made by the parts sales center, such as defective parts that were not checked on delivery or parts that were incorrectly ordered, installed, and later removed. Other problems for the labor sales profit center occur when part orders are not received and, for example, an automobile must be removed from a bay and the job could not be completed. The question is whether these errors and problems should require a transfer of money to the labor sales profit center to cover the entire cost or loss on a repair. Should the labor sales area have to share in the loss when the amount collected by the parts sale area is less than the amount charged for labor by the facility? Should there be a set fee for labor charges made to the parts sales profit center regardless of the amount of the reimbursement or amount the facility charges for labor? If the parts specialist is paid a percentage of the net income/profit made by the parts sales profit center, the amount of the payment will directly impact his/her take home pay.

Finally, the greatest advancement in the parts sales area (mentioned earlier) is the use of the computer for checking on the cost of parts for estimates, checking on their availability, and ordering parts. When preparing an estimate, for example, the cost of the part can be downloaded from a supplier or suppliers into the computer (without talking to anyone on the phone) and the computer program can instantly markup the cost of the part according to the markup strategy of the facility (or the markup can be changed and manually inserted by the specialist). Then the computer can present the labor time recommended for the job and, upon command, compute the labor charges for the repair (time for the repair times the labor charge of the facility). Next, the parts specialist can instruct the computer to add other costs, such as a charge for shop supplies based on the total cost of the parts and labor. Finally, the computer can compute the sales tax to gain a total charge for the repair.

After an estimate is prepared, the parts specialist can forward it to the computer at the service consultant's station where it can be printed out and shown to the customer. If approved by the customer, the computer can print out the work order for signiture by the customer with a hard copy for the technician. The service consultant can send the approval back to the computer of the parts specialist who can send the computer order to the parts supplier.

Review Questions

1. What is a profit center?
2. What is the responsibility of a service manager to the labor and parts sale area profit centers?
3. What is the format and purpose of the parts sales management worksheet?
4. How are the target metrics for the line item accounts used on the parts sales management worksheet?

PART III

PRACTICAL EXERCISE

Small Group Breakout Exercises

1. To manage the financial operations of the service department, the CFO uses a traditional income/expense account discussed in Chapter 10. The CFO generates a monthly report similar to the one found in Figure 10-4 and expects you to monitor your financial operations from the statement he produces each month. The CFO realizes, however, that there are more line items than are realistically needed by your department. Prepare a report for the CFO that identifies the account line items you could use to monitor your operations and improve the service department. Support your recommendation with appropriate textbook chapter citations.
2. Although the CFO uses a traditional accounting statement, you prefer to use a labor sales worksheet, such as the one shown in Chapter 11. In a report to the CFO, describe the managerial accounting worksheet and explain why it is different from the traditional accounting statement; for example, point out the type of information it provides and how it can be used to improve the service department's operations. Support your answers with citations from the appropriate textbook chapters.

3. The procedures at the facility require all receipts and deposits to be handled by the business office. The part-time cashier has quit her job and the bookkeeper is suffering from a serious illness. The CFO has informed you to be prepared to handle all payments, bank deposits, and daily reports plus petty cash for the service and parts department. Prepare a set of procedures for the CFO that explains how you will assume these responsibilities, mindful that you must maintain and guarantee internal control over all income transactions and the protection of company resources. Support your answers with citations from the appropriate textbook chapters.

4. Unfortunately, the Parts Specialist had to be terminated. After his departure, you checked his records and accounts and realized that the markups expected by CFO were not calculated properly. In your conversation with the CFO, he decided you would use a variable markup to price expensive parts more competitively. Prepare a proposal for the use of variable markups with a justification for your approach. Support your answers with citations from the appropriate textbook chapters.

5. The CEO and CFO decided to divide the current service department into separate profit centers at the recommendation of the manufacturer's representative. Your title will now be the Director of Service and Parts and you will oversee a manager of the service department and a manager of the parts department. You are directed by the CFO to view your operation as three profit centers. One profit center will focus on maintenance and light repairs (oil changes, tire rotations, tire replacement, etc.), a second will take care of general and specialized repairs, and the third will be responsible for all parts sales to the service department and the general public (traditional parts department). You must now examine the income/expense statement and assign line item accounts to each profit center. Start this exercise by going back to the income/expense accounts shown in Chapter 10, Figure 10-4, and prepare a suggested list of accounts for each center. The difficulty in this exercise is determining how to split up certain line item expenses so that each profit center pays its fair share and absorbs the proper amount of expenses. You must explain how you can split up expenses into the appropriate percentages for each profit center and provide a rationale for your decision.

6. The five-bay independent service facility next door to your service facility is for sale. The owner of the facility has a good reputation and your CEO and CFO think it may be an opportunity to expand the service department. Their idea would be to buy the service facility (the building is not for sale) and put the light-maintenance and repair profit center (oil changes, tire rotation, tire replacement, etc.) in that location. The dealership service bays would then focus on complex repairs. Based on the information presented in 6A and 6B they want you to first use your managerial worksheet to determine whether the

business is profitable as is, and if the calculated percentages fall within the target metric. Next, they want you to identify changes you could make to streamline the operation to meet the needs of your dealership. These changes should then be used to rerun the management worksheet; however, redundant expenses paid by the dealership (such as information system fees, advertising, and development programs, among others) should not be included in the worksheet.

A. Start the analysis by putting the information obtained from the owner of the independent service facility (given next) into the spreadsheet found on the Delmar Website (http://www. autoed. com/resources/supplemental/index.aspx). Then conduct your analysis. The numbers you need to calculate labor sales are a labor rate of $50 per hour and the number of labor hours sold each month equals 688 hours (all sales were at the stated labor rate, none of the sales were at a discount, and there aren't any transfers from the parts sales area). The entire 688 hours is generated each month by four flat-rate technicians who are paid $15 per hour (each earns/produces 172 hours per month). The owner of the facility is also the service manager and pays himself $4,000 per month (he does not receive a commission or bonus). There is a service consultant who is paid $1,500 per month plus 5 percent of the gross labor sales (the month studied would be 5 percent of 688 hours x $50 per hour). Uniforms cost $300 per month and there are no retirement or medical benefits. The owner sets aside $500 for development programs (training) each month. Insurance costs $1,500 per month and there is one equipment lease for an alignment machine and rack at $1,495 per month. Telephone charges equal $600 per month and a Yellow Pages advertisement costs $450 per month plus another $1,200 is spent on other forms of advertisements and promotions. The facility spends $400 on special tools and $200 for garbage removal each month. The rent for the 3,000-square-foot building is $2,000 a month plus maintenance costs of $500, taxes at $400, and utility bills of $375. Finally, the bank fees are $295, office supplies are $200, and information management fees (technical information) are $300 per month. Other expenses were not given.

B. The parts area generates $19,000 a month and the owner does not have any breakdown of how the income is earned. He pays the parts stores around $9,000 per month and doesn't use the discount from 2 percent/10 net 30. Sublet repairs cost around $2,000 per month and he figures tire disposal at $800 but does not have any other cost of parts figured out. The owner hired a helper to handle parts orders, returns, and so on. He pays this person $1,375 per month and supplies him with a uniform at $50 per month. The owner explains that he simply plans that all expenses (phone, etc.) are paid from labor sales without anything charged to the sale of parts with the exception of waste removal, which is $300 per month.

PART IV

FINANCIAL ANALYSIS

CHAPTER 14

DIAGNOSING FINANCIAL OPERATIONS

LEARNING OBJECTIVES:

Upon reading this chapter, students should be able to:

- *Explain the purpose of a financial diagnosis.*
- *Describe a quantitative diagnosis and give an example.*
- *Describe a qualitative diagnosis and give an example.*
- *Explain the purpose of financial modeling.*
- *Describe the difference between variable and fixed costs.*
- *Prepare a breakeven analysis.*
- *Present an example of a control chart.*

Introduction

Labor and parts sales make up the business core of an automobile service facility. If a service manager follows a well-defined business plan, has hired good people, and sees that customers are satisfied, then the facility should be financially successful. This is important to a service manager, because regardless of how popular and well intended he or she may be, an owner cannot keep a manager if the service facility is not a financial success. Therefore, the service manager must constantly monitor and diagnose the financial performances of the facility, preferably through the diagnosis of its profit centers. Because a service facility can be a complex organization, the use of profit centers as discussed in the previous chapters is recommended.

The previous chapters described the financial information available to service managers from worksheets and an income statement, and then what the information represents. This chapter concerns the use of this information to conduct a financial diagnosis of profit centers and the facility. If the findings of a diagnosis appear to be dramatically unacceptable as described later, the next chapter presents a structured process to be followed to conduct in-depth assessments. Therefore, the purpose of this chapter is to introduce the basic diagnostic tools and principles that service managers should use in a financial analysis, how to calculate the breakeven point for a profit center and facility, and then two methods to track performances.

Diagnostic Procedures

Quantitative Methods, Management Science, and Operations Research are used interchangeably when describing an analysis of an organization, such as a service facility. *Management science,* however, proposes that a scientific approach is utilized to study an organization. **Operations research**, by contrast, regards the examination of processes, that is, the procedures followed by a service facility to make repairs on automobiles. *Quantitative analysis* means numerical measures are used to provide mathematical responses and conclusions that are often used in decision making. This chapter focuses on a few of these basic techniques that a service manager may use to diagnose operations. The discussions do not include higher-level statistics or probability theory. This is not to imply, however, that they should not and cannot be used but, rather, that they are beyond the scope of this text.

Access to multiple sources of up-to-date and accurate quantitative data (traditional and nontraditional accounting information) presents the service manager with an opportunity to conduct operations research designed to carry out performance reviews, track changes, and predict trends. The conclusions based on the findings from the operations research inquiries should serve to identify problems. Using the scientific management approach to diagnose and solve the problem, the manager needs to:

- Clearly define the problem.
- Prepare objectives or questions that will solve the problem.
- Collect the data needed to respond to the objectives or questions.
- Analyze of the data.
- Propose options and come to a conclusion.

The scientific approach requires the recognition of the content discussed in the first part of the text, which covered the preparation of a strategic business plan. As explained in these chapters, this plan includes goals with specific objectives stating what the service facility will accomplish. Simply defining a problem is not enough. When diagnosing a problem, the service manager must always keep in mind the strategic business plan's visionary goals and measurable objectives. In other words, assume that a problem is identified and the service manager's conclusion is that the facility needs additional repair bays. If one of the objectives in the strategic business plan is to increase sales with existing resources, this is not a viable recommendation for the owners. Rather, the service manager must arrive at another recommendation or attempt to drive change into the strategic business plan.

Data Sources

Some managers may believe the income statement (traditional accounting information) provides sufficient data to conduct an analysis and diagnosis for making all management decisions. Although an income statement contains important information, such as sales, cost of labor, and profit (the primary target of an objective), for a quantitative analysis it does not contain enough information to conduct operations research. In addition, the income statement is not designed to present information needed to monitor daily or weekly performances. Consequently, although an income statement is a valuable source of data, service managers often must come up with additional information (step 3 above) to make management decisions.

An example of information outside of a traditional income statement and available to a service manager would be technician flat-rate performances. Assume the problem is that the profit objectives of the company are not being met. When the technician performances are collected (step 3 above), the service manager can set proposed performance target levels (objectives) to compare their effect on profits (step 4). From the analysis, the manager may conclude that, when technician flat-rate performances are too low, the profit objectives of the business cannot be met. Options can be explored by the service manager and owners, and one would be chosen for implementation. For example, based on the option selected, the conclusion may be to increase the number of technician flat-rate hours by 11 percent (this should not be an arbitrary number but should be obtained from worksheet calculations). The service manager must then determine how he/she will achieve this target level.

Of course, the above example is oversimplified. In reality, such a diagnosis would consider a number of other questions and objectives to be investigated. For example, these might require the collection of data on sales, customer service satisfaction, advertising, job work orders, use of shop space, type of repair work conducted, and so on. As some would say, the devil is in the details and so, in response, the service manager may have to collect qualitative data to give the investigation more detailed information. In other words, to thoroughly diagnose a problem, quantitative data as well as qualitative information is usually required.

The diagnostic work of managers is not unlike that conducted by automobile technicians when they obtain scan tool data and determine that a system or part is not working properly. Technicians come to a conclusion by comparing the data retrieved from the scan tool to the targets needed for proper performance. At some point, however, the technician must open the hood to look at and further test (using other tools) the systems or part that is suspected to be at fault. Therefore, as with scan tool data, financial data can take a manager so far but sooner or later the manager must open the hood to investigate further what is not working properly. This may require additional quantitative data as well as qualitative data to come to a conclusion and make a decision.

Diagnostic Tool: Quantitative Blueprint

One diagnostic technique for service managers is to prepare a financial model that serves as a blueprint to follow. The blueprint should indicate how a service model should look and operate. For example, blueprints were presented in Chapters 11 (labor) and 13 (parts) to exhibit how data (line item accounts in the sections on the worksheet) relate to each other. The service manager must conduct an analysis of the data to determine how the pieces interact with each other and how they affect the net income (loss). Naturally, a worksheet is only useful if it actually presents data that can be diagnosed to indicate whether the business mission, goals, and objectives are being supported.

In many cases, the analysis of the data provides a diagnosis that offers alternative solutions. Neither of the alternatives may be a wrong decision and so the challenge is to pick the best alternative. For example, if additional repair work is needed, the alternative may be to pay the technicians overtime or hire more technicians. Using the worksheets, each alternative can be examined by what is known as modeling. Naturally, realistic numbers must be entered into the worksheet for each alternative so that a hypothetical output (net income) can be obtained and a manager can look at the effects of each outcome to chose the best solution.

When a quantitative model accurately portrays how a service facility functions financially, service managers may rely on current data, such as traditional accounting information (gross profit) and nontraditional accounting information (flat-rate hour production per day). A diagnosis of the model's quantitative analysis is composed of the following two parts: the model and the analysis.

1. *Model:* The model consists of current quantitative line item income and expenditure inputs to create a baseline. Note that after the model is created, the line item income and expense amounts can be changed (using an Excel program to automatically calculate the changes) to find the output of different options. The model's best output can eventually become a financial blueprint for the service manager to follow.

A good model that accurately reflects a service facility's performance may be used to evaluate how various scenarios affect the service facility's financial output and targets. For example, if heating oil used to heat the service facility increased by $.50 a gallon over the past year, how much must the labor rate increase? In other words, when a model is available, the manipulation of the variables (in this case is the heating oil cost) can help the manager make a decision (in this case a labor rate adjustment) to preserve profits.

2. *Analysis:* The data obtained from the model can use mathematical equations to present optimal operating ranges. Specifically, a manager may want to know at what sales level (or flat-rate hours) the profit equals 0 (no gain, no loss), which is referred to as *breakeven.*

Diagnostic Tool: Qualitative Data

To aide in the diagnostic process, qualitative data may be effectively used. Qualitative information consists of subjective responses as opposed to direct quantitative measures. These subjective responses may be about people's attitudes, beliefs, opinions, feeling, thoughts, satisfactions, and so on, such as responses to a customer satisfaction survey. Needless to say, this information should not be overlooked as it often provides invaluable insights when combined with quantitative findings. To help in the analysis of qualitative data, it may be presented in a quantitative format. For example, customers may be asked to rate their satisfaction with the service facility's performance on a scale of 1 to 10, with 1 being the lowest rating and 10 being the highest. Although the measure of customer satisfaction is qualitative, the scale (called a Likert scale) makes it possible to use mathematical and statistical calculations to draw conclusions. Therefore, when possible, qualitative assessments should be constructed so the information is presented in a numerical format in order to use it properly in an analysis.

Qualitative information is important to the decision-making process. More specifically, qualitative information may either support or reject the quantitative findings. For example, if higher oil prices are driving the decision to raise the labor rate but the shop furnace is old and inefficient (black smoke is billowing from the chimney), it might be a better idea to look at alternatives such as a new furnace or maybe a waste-oil furnace. In the analysis process, if furnace replacement makes more sense so that less oil is used, then replacement of the furnace would be part of a capital improvement. Unlike an expense, such as buying heating oil that would be recovered by increasing the labor rate, replacement of equipment is a capital investment by the owners. Equipment replacement should be part of the strategic business plan and, as such, money should have been already be set aside from past profits (or a plan already in place to borrow or obtain the money) to pay for it.

Another example of the importance of qualitative information in decision making would be whether or not a facility should hire more technicians. From a quantitative point of view, hiring more technicians might initially be the best decision because the analysis projects a greater net profit. However, the decision to hire more technicians requires more than just good numbers. Others in the organization will need to buy in to the decision, such as flat-rate technicians, whose pay may be affected, owners, who will bear the risk of higher expenses if sales do not increase as projected, and even customers, who may not like the busier shop environment. Each source is important to consider and the qualitative information gathered from them should not be disregarded just because the quantitative analysis portion meets a specified objective or solves a problem. This presents a new dynamic for some service managers because they are prone to use either a qualitative or quantitative approach to decision making and few use both.

Therefore, some decisions should be made by a management team or focus group whose members can offer different perspectives. Some members may be qualitatively oriented and some quantitatively oriented. Although not all decisions can be made in this way, to take this a step further, when a tactical plan (see Chapter 3) is developed or a portion of the tactical and operation plan is reviewed to improve some aspect of the business, a management team approach would examine the strategy from various perspectives. Specifically, the team must examine whether the qualitative information (interviews with various people, the environmental influences such as business conditions, area trends, among others) compares positively or negatively with the quantitative data. This comparison of information increases the probability that the best option or plan is chosen.

Financial Modeling

The labor sales worksheets (presented in Chapter 11) and parts sales worksheet (presented in Chapter 13) are financial models designed to give a service manager the bare bones financial information about two profit centers in the service facility. Although the income statement and balance sheet are important documents for a manager to understand and use, they contain more information than what a service manager will need for planning and analysis purposes. For example, an income statement contains line items (such as depreciation expense) and, although it is important to understand, its primary use is to reduce the net income for tax purposes. Therefore, a worksheet is similar to the income statement; however, it contains fewer line items that are specific to the profit center. Internal users such as the manager, owners, and consultants use the worksheets in conjunction with other information to review and set labor rates and answer basic questions such as: How many flat-rate (or billable) hours must be generated each week to reach a breakeven point? Using this approach, a manager can examine different markups and determine how a breakeven point changes.

Financial worksheets are models that monitor the performance of each profit center. How often a worksheet is updated depends on the situation but at a minimum it should be tracked from month to month. Answers should come from careful examination of the worksheet and should include how a profit center's income and the gross profit changed for the past month. Although the number itself is important, typically the percentage is of greater significance because it shows how the cost of sales to income increased, stayed the same, or decreased over the past month. Depending on the percentage, a manager could gain a better idea about the degree of change. For example, if the gross profit for the parts sale area dropped 5 percent or more, the parts markup used by the parts specialist may need to be examined.

Target Metrics: A Model within a Model

Another type of model presented within the financial worksheets was a target metric. Target metrics are optimum operating tolerances (numerical ranges, specific numbers, or percentages) within which each line item on the worksheet should operate. Each month, a manager should review each line item to compare it to the prior month and also to the target metric when given. The questions to be answered is whether the service facility performance on each line item met its target metric. When a service facility does not meet a target metric, then an examination of the prior month's performance should be examined relative to the current month to determine whether there was a change. Ideally, when a profit center's target is met, its performance is acceptable and should be financially successful. If not, the service manager must investigate by collecting additional data, possibly qualitative information, diagnose the findings, and come to a conclusion with options to correct the problem.

Breakeven Model: Variable and Fixed Costs

Breakeven occurs when the income of the facility or profit center equals the costs and expenses to provide a zero net income. The calculation of a breakeven model starts with the separation of the line items from the most recent worksheet(s) or even the annual income statement into three areas:

1. sales (income for the profit center or entire service facility if desired).
2. variable costs (costs and expenses that change with every unit of production, such as flat-rate technician wages).
3. fixed costs (costs that do not change each workday with production changes such as the wages of hourly technicians).

When the financial information is obtained from a yearly statement (such as an income statement) the sales and cost totals are divided by 12 to obtain monthly averages to calculate the breakeven. A longer period of time is desirable to get a more accurate portrayal of the sales and costs (this helps avoid fluctuations that normally occur with seasonal demand among other reasons). Although a breakeven can be found for a period of time such as a year, meeting the average breakeven amounts each month, of course, is not likely because of the irregular and sometimes unpredictable level of monthly business activity. This uneven financial activity requires a *Volume-Based Allocation (VBA)* plan, discussed in Chapter 15; however, for the discussion and application of a breakeven analysis, the model uses the average monthly amounts.

To calculate the breakeven for a labor sales profit center the accurate separation for the calculation of the difference between variable and fixed costs is key. For our example, assume that:

Labor sales equals an average of $20,000 per month

- variable costs (cost of sales and expenses such as salaries of flat-rate technicians, management bonuses, any expenses that change with the volume of production) equals an average of $8,000 per month
- fixed costs (overhead and any other costs that are constant regardless of the number of sales made, such as management salaries, hourly paid employees) equals an average of $8,000.

Figure 14-1 shows that when the variable costs (VC) are subtracted from sales, the difference is known as the contribution margin (CM). In the example, the contribution margin equals $12,000 or 60 percent of sales. When the fixed costs (FC) are deducted from CM, the net income (NI) is $4,000.

Sales	$20,000	100%	$50/FRH
Variable Cost (VC)	$ 8,000	40%	$20/FRH
Contribution Margin (CM)	$12,000	60%	$30/FRH
Fixed Costs (FC)	$ 8,000		
Net Income (NI)	$ 4,000		

FIGURE 14-1 Average monthly performance for labor sales.

In the next part of the analysis (see Figure 14-1), assume the labor rate of the service facility is $50 per flat-rate hour (FRH). If 40 percent of each sale goes to the payment of variable costs, then $20 of the $50 received for each hour sold pays the variable costs while $30 per hour remains for the contribution margin.

Once the basic equations (Sales − VC = CM and CM − FC = NI) and the percentages are computed, the service manager can calculate the breakeven point for flat-rate hours and sales. To make the breakeven analysis relevant, then 400 hours are generated each month based on the sales of $20,000 ($20,000/$50). Therefore, breakeven in terms of flat-rate hours and dollars is as follows:

Breakeven on the Sale of FRH:
FC/CM = $8,000/$30 = 266 FRH.
The breakeven for the sale of labor would be:
FC/CM% = $8,000/.6 = $13,333.

Consequently, the facility would have to sell 266 flat-rate hours in a month for $13,333 before $1 of net income is earned. To illustrate that this is true, the mathematical proof is shown in Figure 14-2.

For a service manager, once the breakeven is reached, the rate at which profit is earned increases rapidly. Specifically, once sales reach the breakeven of $13,333, then $3 (60 percent) out of every $5 is profit with $2 (40 percent) going to pay variable costs. This is an important reason why fixed costs must be kept within a target metric tolerance. Too often owners and managers think fixed costs do not have a great meaning; however, they must realize that money is not made until the fixed cost are covered.

As owners, the authors monitored the sale of labor and parts to gauge when their breakeven would be reached or if it would not be met by the end of the month. Needless to say, understanding breakeven point influenced many decisions. For instance, if sales fluctuate greatly and are generally lower than expected, or even declining, a management response might be to make as many fixed costs as possible variable costs. This would mean moving hourly paid technicians to flat rate, eliminating salaried management positions, and paying service consultants a percentage of the sales and managers by a bonus system. Because the variable costs would be directly related to the amount of sales (as opposed to fixed

Sales	$13,333	100%	$50/FRH
Variable Cost (VC)	$ 5,333	40%	$20/FRH
Contribution Margin (CM)	$ 8,000	60%	$30/FRH
Fixed Costs (FC)	$ 8,000		
Net Income (NI)	$ -0-		

FIGURE 14-2 Average monthly performance for breakeven on labor sales.

costs), the breakeven point would be lowered. In addition, it would hopefully provide an incentive for personnel to help increase sales.

In contrast, when sales are up (business volume is steady and perhaps growing), the breakeven will be reached earlier in the month than usual. In this case, management may want to maximize profits by increasing fixed costs as opposed to variable costs. Technicians should be paid by the hour (not flat-rate) and additional employees (to help get more work done) should be paid as a fixed cost (hourly rates). The service facility would now operate more like a like a factory because the growth in sales and production would be covered by fixed costs (provided the workers are competent and can efficiently get the work done). This would allow more of the increase in sales to become net income. As a compromise between these two tactical approaches, some facilities offer a blend of fixed and variable costs. For example, technicians are given a modest hourly wage guarantee with a flat-rate wage (either pay or an incentive added on) kicking in when a certain point (breakeven) in sales is reached.

Model Projections and the What If Game

Using a computer spreadsheet such as Excel allows the worksheets or breakeven model to help create projections. This allows service managers to play the what if game. A quantitative what if question would be: If the labor rate is increased $2, what will be the change in net income assuming costs remain the same? A qualitative question that would go along with this idea would be: If the labor rate is increased $2, will the facility lose (or maybe gain) some customers (meaning the sales would decrease)? A follow-up question would be: If the facility would lose 5 percent of its customers due to the increase in the labor charge, what will be the effect on net income assuming all costs and expenses remain the same? What if the facility would lose 8 percent?

In other words, before any management decision is made to change expenses, wages, fees charged, markups, services offered, purchase of equipment, method of paying technicians, and so on, the service manager should run a model. This will help the service manager see the effects of the changes and reduce the risk of making a wrong decision. In some cases, the service manager will see the effects of each and help him/her to make the best decision.

In some cases, although a suggestion may seem logical, the financial projections may indicate it is impossible. For example, a service facility may plan to purchase an alignment machine. Before the purchase is made, a worksheet should be prepared for an alignment profit center. This worksheet would include the line items for each of the four sections (income, cost of sales, expenses, and net income). As discussed in Chapter 13, some expenses will be split with other cost centers so accurate estimated amounts are entered for each of the line items.

Once the worksheet has been created, sales, fixed costs, and variable costs are identified so breakeven (as describe above) can be calculated. The outcome, of course, is to see the impact from the increase in fixed costs (monthly payments for the machine). This also gives the manager the ability to play with different scenarios so that the price of the alignment and volume that must be performed can be balanced. The reason this is important is because as specialized equipment (alignment machines/racks, diagnostic equipment, emission equipment, among other items) becomes much more expensive, it may not be possible to cover the cost with an hourly labor rate. Instead, flat fees may be necessary to cover the costs associated with the profit center, especially when the volume of business for the profit center has a theoretical limit. For example, an alignment bay may have a theoretical limit of eight alignments per day with a practical limit of six. Of course, qualitative information also will have to be collected; for example, if the facility also sells tires, what will be the influence on the sale of tires (and warranty return for worn tires) if alignments are not offered. If the alignment profit center never reaches its projections, what other kinds of services could the bay be used for? Anytime a service facility plans to add a new service, such as alignments, or if the sale of an item, such as tires, is expected to increase and use more resources (space, capital, and employee time), then a profit center should be created or in some cases combined with the another profit center (for instance, alignments and tires might be combined into one profit center). Whether a profit center is created and how it is placed within an existing business requires the service manager to create and run several what if models.

Tracking Performance through Control Charts

Control charts are based on a breakeven analysis and are another way to help a manager see what is happening in the service facility. They are like a worksheet because they can be used to monitor business operations on a daily, weekly, monthly, or annual basis. They can be created for the entire service facility but are most useful when they monitor the performance of different profit centers and income statement line item accounts and summaries, such as total sales, sales of labor, sale of parts, cost of labor and goods sold, gross profit, total expenses, line item expenses, net income or profit, variable costs, fixed costs, marginal contribution, technician flat-rate hours, and so on. When a spreadsheet program such as Excel is used, control charts can be created from the graphing function in the software.

The advantage of using control charts is that they permit managers to visually track small movements (daily, weekly, and monthly performances), so the manager can pick out trends and see the big picture. Being able to see fluctuations in performances and recognize trends as they develop can speed up decisions and expedite actions that must be taken (such as hiring or laying off personnel, increases in advertising, and so on). In other words, time is of the essence in some situations and failure to recognize a trend early enough can result in losses that cannot be easily rectified in the short term.

To plot data on a chart, horizontal and vertical lines are used. The horizontal line typically represents time while the perpendicular line usually signifies the data collected, such as the weekly (horizontal line) dollar amount (perpendicular line) of sales. When a control chart is used, at least three horizontal lines are added to a chart. The lower line in the middle represents the *lower specification limit,* which would be breakeven, such as the breakeven number of flat-rate hours to be generated each week by the technicians. For example, from the breakeven analysis discussed above, the breakeven FRH needed per month was 266 hours. Since the control chart is to record weekly performances, the 266 breakeven monthly hours would equal 62 breakeven hours per week (266/4.3 = 62 hours per week). Therefore, the lower horizontal line on the control chart showing technician labor hour production would be at 62 hours, as shown in Figure 14-3.

The middle line on the control chart represents the *target specification,* such as the number of labor hours that should be produced. In Figure14-3, the target is 400 FRH per month or 93 FRH per week (400/4.3 = 93 hours per week). This target represents the production needed to generate the profit expected in the strategic business plan prepared by the owners. Finally, the maximum amount of work the technicians can handle *(upper specification limit)* is 124 hours. Upper limits should be set because when technicians, or any person doing a job, reach a maximum level of production, their efficiency and effectiveness begin to fade and there is a greater chance for errors and accidents.

After the control chart is prepared, the manager can record and track weekly performances. For example, as shown in Figure 14-3, the production of FRH shows a downward trend. The first week the target was hit

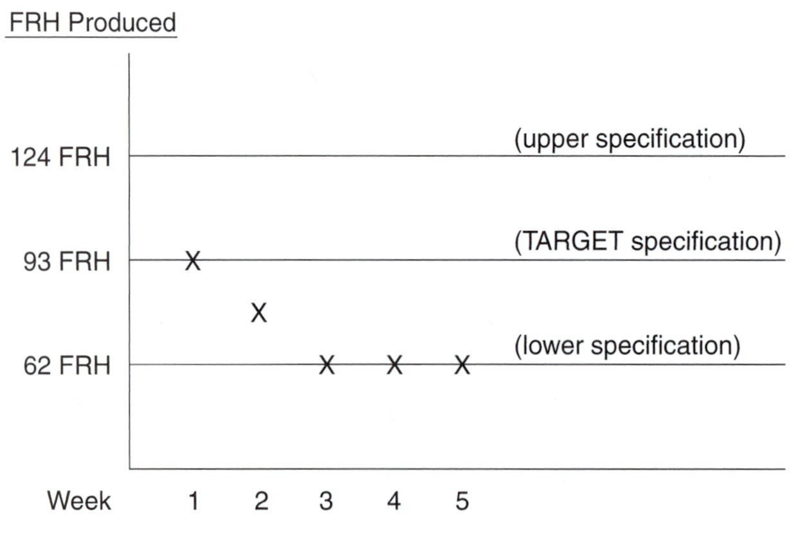

FIGURE 14-3 Flow chart with specification limits.

and then declined the second week until FRH production reached breakeven (no profit or loss) and stayed there for three weeks. The profit target expected by the owners will not be met. The service manager should see that a trend is beginning to develop and plan the actions he/she will take. The service manager does not have a lot of time to put a plan into action. If the manager waits too long, production could go below the lower specification limit. This could mean notifying the owners that the facility may experience a loss, which could be an undesirable experience. Of course, the service manager may do nothing and production could go up; however, the service manager would be taking a risk for the owners as well as a personal risk.

Once the trend is recognized, additional data can be collected to help understand the trend, and options can be examined with the idea of selecting the best option. Based on the example shown in Figure 14-3, the service manager has to improve sales or cut expenses/costs. To cut costs the manager could ask for voluntary (paid and unpaid) vacation time from the employees, cut back working hours, or lay off workers. Additional analysis would reveal that while these actions would help preserve profits (by basically reducing the lower specification line further), this action could cause problems because if customer demand increases, workers will not be available to generate more flat-rate hours. The other action would be to try to reverse the flat-rate hour (FRH) production trend by growing sales. This is where the hard work begins. Advertising, marketing (direct and indirect), and promotions (offerings specials, calling customers who are due to have maintenance service or an inspection, and having technicians check automobiles carefully for additional work) must all come together in a team atmosphere with the idea of getting more work.

In the second control chart shown in Figure 14-4, the data shows a different trend. In this chart there is a fairly strong pattern until week 6, when the upper specification limit is reached. The production holds for another week, backs off, and goes up again. The three-week trend preceeded by two weeks of above breakeven performances suggests the need for expansion. In other words, the facility is working at maximum capacity and how much business is getting away is unknown.

By not taking any action may mean the opportunity to grow and expand will be lost. Possibly, the service manager should recommend to the owners that another technician be hired; however, expansion could be a problem as space may not be available. This would mean another service bay would likely need to be equipped or even added to the building. In addition, expansion means that while the upper specification will go up (so more work can be performed), so too will the breakeven point (lower specification line). In such a case, the trend suggests that the strategic business plan of the facility must be re-evaluated. Why have the flat-rate hours (FRH) increased? Perhaps it is because of the reputation or the facility may be offering services beyond the scope of the mission statement; if so, this may be an opportunity or a threat. The increase also may directly or indirectly mean that some of the business goals and objectives have been met. If this appears to be true, then perhaps new ones need to

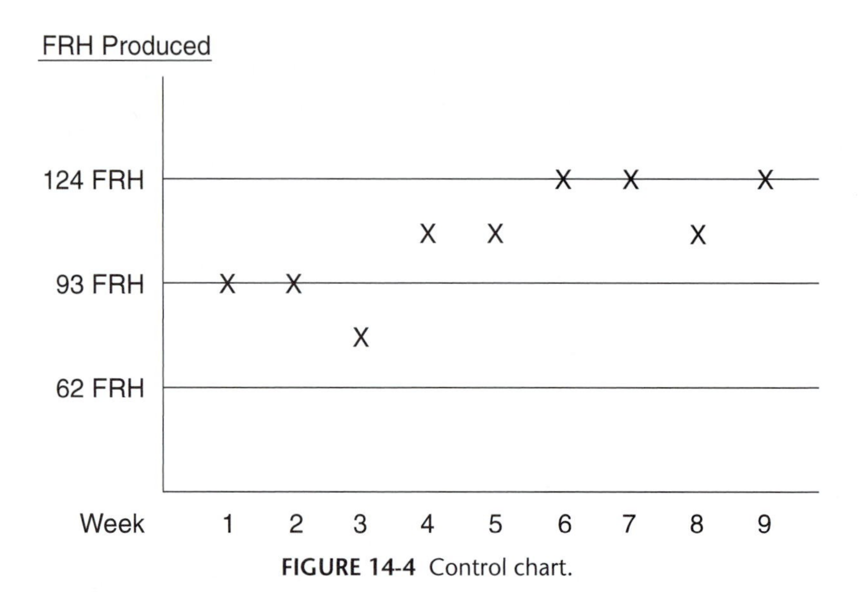

FIGURE 14-4 Control chart.

be created or existing ones expanded. Other questions about the business also are important; for example, what form does expansion take? Does it mean adding space to the facility, extending the working hours (evenings and weekends), or opening another facility?

The answers to all of these questions are part of the analysis process and exploration of available options that ultimately lead to more questions. An important point, however, is that a decision must be made. Making no decision is actually a decision to do nothing.

Red Light, Orange Light, Green Light

To illustrate in more detail the ways in which a control chart can be used even more effectively in a facility, two more limits (horizontal lines) can be inserted into the chart. These additional lines can be placed above and below the target line to represent the area where most data entered should appear. The lines are called the *Upper and Lower Control Limits* and the area between them represents the *Green Zone,* or Go Zone. This is the zone where the manager hopes all data will reside when entered (naturally changes, such as expansion over time, would change the entire chart and green zone as new levels are reached). When trends are shown in this zone, a service manager must have the authority to act and react without the approval from upper management or owners.

When data is shown above the upper control limit line but below the upper specification limit line, it is in an area called the *Upper Orange Zone,* or Upper Caution Zone. Likewise when data is shown under the lower control limit line but above the lower specification limit line, it, too, is in the orange zone, but is referred to as the *Lower Orange Zone,* or Lower Caution Zone. When data and emerging trends are in the orange zone, typically, the service manager is expected to notify and advise upper management and/or owners.

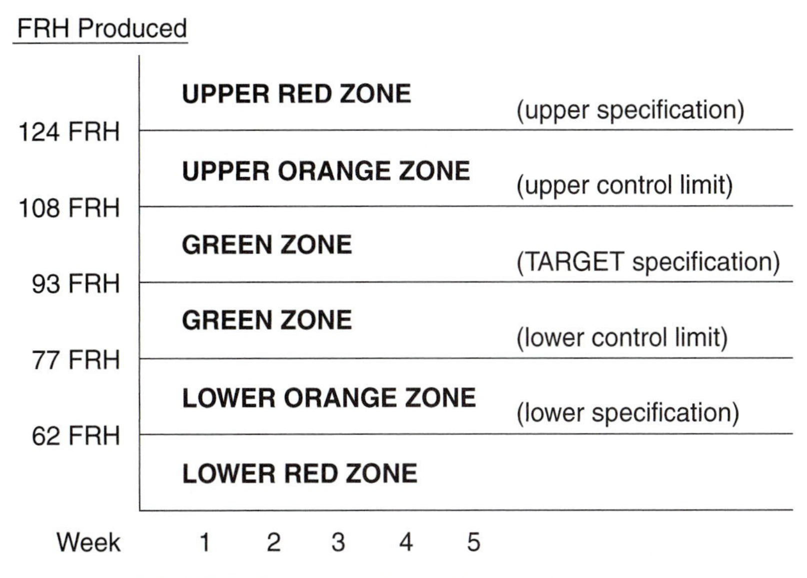

FIGURE 14-5 Control chart with specification limits.

Anytime the data trends appear outside the upper specification limit, they are in the *Upper Red Zone,* or Upper Danger Zone. Data appearing outside the lower specification limit is in the *Lower Red Zone,* or Lower Danger Zone. Data and trends in this area suggest the service manager must provide a full report to upper management and/or the owners for them to become involved. When trends appear near or outside the upper or lower specification limits, upper management or, more likely, the owners must take action. This is because the owners may have to commit capital assets for expansion (upper specification line) or cash to cover losses (lower specification line).

Setting Values for Control Limits

Obviously, the values set for the target, control and specification limits on a control chart are critical. Ideally, the amounts set for the horizontal lines are from statistical calculations that involve complex mathematical equations. Because these statistics are beyond the scope of this text, the authors propose the use of a practical approach (exhibited in Figure 14-6) for the assignment of values to horizontal lines that can be used by all service managers. Note that FRH and technicians have been placed in parentheses because the control chart could be used to plot sales, cost of goods sold, gross profit, net profit, total expenses, and so on. For example, the upper specification is based on the service facility's potential (number of bays and hours open per week), while the upper control limit is based on all of the facility's technicians' potential (flat-rate hours they can generate). The target is the amount to be reached to meet the planned profit stated in the business plan while the lower specification limit is breakeven in this more complex control chart example. The lower control

Control Limit	Assigned Value
Upper Specification Limit	Potential (FRHs) of the Facility
Upper Control Limit	Maximum Efficiency (of Technicians)
Target Specification	Planned Profit per Strategic Business Plan
Lower Control Limit	Minimum Performance Expectation of the Technicians
Lower Specification Limit	Breakeven

FIGURE 14-6 Values assigned to control limits.

limit in this more complex control chart is different than Figure 14-4 and is likely the most difficult because it represents the level where employees do not have enough work to do and someone needs to be sent home early.

Although the spacing between the lines appears even in the examples given, in reality this method may produce unequal spacing between the control chart lines. However, as a practical method, this should not present a problem in terms of analysis.

As a result, recognizing the practical applications of the control charts, they must be carefully designed and the data must be accurate. Errors and ghost readings (meaning the false interpretation of the data) can be very damaging to the facility, to employees, and to the career of the service manager. Although the method illustrated focused on flat-rate hours, other measures important to performance of the service facility can be used. In fact, some managers track this type of information using other methods; however, none are quite as easy to use or to present visually as the control chart. When control charts are compiled as a form of macro analysis over a period of time (weekly recordings are converted to monthly and/or quarterly charts), they offer yet another level of analysis, which will be discussed further in the next chapter.

Closing Thoughts

As discussed in the first part of the text, a strategic business plan is prepared so that employees know what they are to accomplish. The plan contains the goals and objectives to be attained as well as operational procedures to explain how employees are to meet the expectations of the plan. As the second part explains, personnel who can perform the tasks required by the procedures are hired by the facility. This includes the - service manager, who is expected to direct and coordinate the offering of the services to customers. Of course, this assignment includes the management of the money invested to finance the plan as well as the money earned and spent by the business. After the plan is put into practice, the fundamental questions asked of all managers, sometimes on a daily basis, are: How is the game plan working? Are we winning (meaning are you meeting the goals and objectives of the facility and making the expected profit)?

Managers are expected to answer the fundamental questions. They are expected to know how well the facility is performing both in financial terms and nonfinancial terms. Admittedly, some managers are not expected to be able to answer these questions. Their duties are limited to supervision and taking care of problems. Other managers, however, are expected to answer the questions. These managers, therefore, must be able to monitor, diagnose, and adjust operations to meet the expectations of the strategic business plans. This requires them to use models and performance control charts to monitor performances, use proper techniques to analyze and diagnose worksheets, and make the best decisions. They must know where the breakeven points exist and track performances for trends that indicate when the facility is doing well and not doing well. When the observe fluctuations in performances, they must determine where the performances fall within the limits of the control chart. Depending on the trends, managers must know when to make decisions, when to advise their supervisors or owners, and when to report and ask for decisions.

As explained in this chapter, fluctuations are the problems for all service facilities. They may be due to economic shifts or they may be seasonal and, most often, they are not caused by anything someone in the facility has or has not done. Regardless, they are difficult for managers to handle and the directives of the strategic business plan do not offer answers. Owners and managers often have feelings of hopelessness and take a position that they just have to wait out the storm. These feelings are not altogether justified. Managers should follow a process to diagnose the problems and come up with decisions or suggestions to can help control, or at least lessen, the influence of these unpredictable and often irregular shifts of the facility. This process is discussed in Chapter 15.

Review Questions

1. What is the purpose of a financial diagnosis?
2. What is quantitative diagnosis? Give an example.
3. What is qualitative diagnosis? Give an example.
4. What is the purpose of financial modeling?
5. What is the difference between variable and fixed costs?
6. What is a control chart? Give an example of the use of performance zones.
7. Describe the purpose of the steps in the scientific management approach.

CHAPTER 15

PERFORMING IN THE LOWER ORANGE AND RED ZONES

LEARNING OBJECTIVES:

Upon reading this chapter, students should be able to:

- *Outline the process to be followed to diagnose a facility when performances are recorded in the lower orange and red zones.*
- *Describe the difference between traditional and nontraditional financial data.*
- *Explain the purpose of a unit analysis and a volume analysis.*
- *Describe the difference between a trend and static analysis.*
- *Present the steps in the process to be followed when conducting an in-depth review of a performance problem.*
- *Prepare a plan for the collection of traditional and nontraditional financial information for an in-depth review of an independent service facility.*

Introduction

As discussed in Chapter 14, a service manager must pay particular attention when performance measures are in the lower orange and red zones of a control chart. Although problems also can be indicated when performance measures are found in the upper orange and red zones, they are quite different from those found when measures are in the lower zones. Although the process and methods to examine the problems are the same, greater distress obviously exists when performances enter the lower zones. A serious slump or extended decline in financial operations often cause service managers to want to do something quickly, which results in a shotgun approach. This is because of the misconception that trying anything and everything will turn a business around. This panic method, however, is a sign the service manager does not know how to proceed to identify and solve the problem. Instead of using the shotgun approach, the service manager must follow a diagnostic strategy similar to the process discussed in Chapter 14. This strategy is not unusual in the automotive business because it is followed to identify and repair an automobile's mechanical problem. The purpose of this chapter is to describe the process and techniques to use when performance declines pose a serious threat to the financial stability, and possibly the survival, of a company.

Chapter 14 presented the basic principles of analysis, diagnostic procedures, models, and control charts, and discussed the monitoring of service facility performances. In particular, the chapter stressed that a manager should always try to stay in the green zone of a control chart. While the techniques used in Chapter 14 focused on diagnostic methods that would help a service manager stay in the green zone, there are times when a facility will enter the orange and red zones. As a result, this chapter focuses on the use of the scientific method discussed in the last chapter to conduct a thorough analysis when performances enter the orange and red zones. In describing the recommended process, the chapter assumes the facility has set up profit centers, as described in the previous chapters. Although operational reviews would include an examination of both financial and nonfinancial data (employee performances, customer satisfaction, changes in environment, and so on), this chapter is limited to a facility's financial performances. The financial information obtained for the diagnosis comes from both traditional accounting information (income statement) and nontraditional financial information (such as flat-rate hour production per day).

The Review Process

As explained in the last chapter, the scientific diagnostic approach requires the service manager to:

- Clearly define the problem.
- Prepare questions to solve the problem.
- Collect data needed to respond to the questions.
- Analyze the data to prepare optional answers to the questions.
- Select an option to come to a conclusion.

This approach should be followed when performances have entered the orange or red zones. Again, as recommended in the last chapter, when performances leave the green zone and enter the orange zone, the service manager should notify the owners and/or senior management. When the readings on a chart enter the red zone, the service manager should plan to write a report to the owners and upper management that summarizes service facility performance. In the report, the service manager should use his/her experience to offer a diagnosis of the problem that will assist in the exploration of possible solutions.

The Analysis Format

The collection of information for a review of a problem should include quantitative and qualitative data to guarantee that the most important points are examined in detail. The qualitative data would include input from technicians, service consultants, parts specialists, and others about each profit center. The quantitative data would include traditional information gained from the income statement and worksheets as well as nontraditional information collected from other records, such as number of repair orders processed.

Therefore, before an analysis is conducted, the service manager must prepare a plan to collect the data. The plan must recognize that when a service facility's performance enters a lower zone, multiple problems are likely to exist in one or more profit centers. As a result, the review of data categories (such as those described below) should *not* begin at the top of a list and continue toward the bottom in a systematic fashion until a deficiency is found. Rather, a manager must examine all profit centers and as many categories as possible at the same time. Unfortunately, because time is usually limited, a manager may be tempted to end the review and reach a conclusion before all of the data has been collected and reviewed. This is a mistake. A manager should not stop a review until all relevant data from all traditional and nontraditional sources is diagnosed. To quit as soon as a deficiency is found is often the reason why a problem is never really solved. Although time is of the essence, a premature explanation that misses the real cause of the problem will only solve a symptom and be, at best, a temporary fix.

The intent of the scientific approach and the related techniques is to dissect the financial performances of the company when it enters the orange and red zones (both upper and lower). The format is not to determine whether a problem exists. This is already known because performances are in a lower or upper zone. To begin the financial dissection, the service manager must collect data for two types of investigations. One is for a *unit analysis,* which is an examination of the facility's account categories and profit centers. The other is for a *volume analysis,* which concentrates on the outputs of the facility and profit centers.

Conducting a Unit Analysis

The unit analysis of a facility should be divided into a *unit-income analysis* and a *unit-cost analysis.* An example of a unit-income analysis would be the collection and comparison of income, gross profit, and net profit data for each profit center. To help guide the process, a question a manager may pose is whether each profit center is generating an adequate income to meet its planned profits. If the answer to this question is no, then other questions must be asked: Did a profit center's labor sales and/or gross profit decline? Did a center's sale of parts decline? If the sales in a profit center dropped significantly, what were the declines in specific parts sales in the center?

For example, a facility may show a serious income decline for the last period and the review of the profit centers found that sales in the air-conditioning services profit center dropped into the red zone. The examination of the sales drop for the air-conditioning profit center would depend, of course, on the time of year the report was prepared. If the report was during a period when the demand for air-conditioning services was high, then the question would be whether there was an unusual cold spell during the period. If there was a cold spell, did it cause a sales decline large enough to have an effect on the facility's income to the point that if fell below the target metric and into the lower orange or red zone? A test of this question might be to assume the income problem in the air-conditioning profit center does not exist by making a change in the air-conditioning profit center worksheet. The question then would be whether this change (to eliminate the problem) caused the service facility income to rise into the green zone. If it did not, then other problems exist. This would cause the service manager to broaden his search to the collection of qualitative data.

An example of a nontraditional unit-income analysis would be to look at the performances of each technician in relation to a drop of income into a lower zone. Does one or more of the profit center's technicians have a problem generating an adequate number of flat-rate hours per week? If so, is the lack of production causing the reduction of income for the facility? To determine whether this is possible, past data collected on repair orders should be obtained. The difference between the past and current incomes in relation to the past and current flat-rate hours of the profit center must be compared to determine whether the decline in flat rate hours caused the dramatic income drop into the lower zone.

In this example, assume the drop in labor sales is not great enough to cause the income problem and a large drop in parts sales is recognized. The data collection process would now shift to the parts sold on each of the profit center's jobs. The manager must look at the percentage of parts sold compared to labor sales since the dollar amount of the parts sold will normally decline when labor sales decline. Upon review of the percentages, the manager may find that the percentage of parts to labor sold declined from one period to another, indicating that parts were not being sold to customers. This would then require the manager to make follow-up qualitative inquiries. For example, qualitative feedback may reveal that some customers were permitted to bring in their own parts as opposed to buying them from the facility. This would naturally cause percentage of parts sales to drop for the profit center. This could even lead to the facility providing extra free labor to solve parts problems (poor quality parts that do not fit properly or are wrong) and cause a drop in labor sales as well.

A unit-cost analysis also uses traditional and nontraditional data to examine the facility and profit centers but from a cost perspective. A problem may not be caused by a decline in sales (income) but by the cost of conducting business. In a traditional unit-cost analysis of each profit center, the service manager might compare accounting data on the actual dollar amounts of costs and expenses and their percentages to income, such as the actual cost of technician salaries and parts, supplies expenses, and fixed costs to income. For example, if the percentage of the labor costs to the income earned by the air-conditioning profit center is much higher than the percentage of the alignment profit center, then the manager must determine why. In addition, the cost analysis would include a comparison of past to current costs and expenses for each center as well as against any target metrics (see Chapters 11 and 13). An example of a nontraditional cost analysis would be to look at individual repair orders in relation to the price charged to customers. For a facility with a parts profit center, the markups would be checked to ensure the sale prices of parts are properly calculated.

Volume Analysis

The sales and income volumes of all businesses, including service facilities, are never constant. The amount of business conducted increases and decreases at different times of the year, different times of the month, and different days of the week. For example, at the authors' facility sales always declined for the month of August and February, the last 10 days in November, the last two weeks in December, and Mondays were not usually great. Therefore, a volume analysis should first try to record income and cost changes of the facility for different time periods. Specifically, sales volumes should be recorded to determine when they were at their highs and lows.

When fluctuations in sales or income are known to occur on a scheduled or regular basis, then service managers should be able to predict changes in volume. With this information, numerous questions can be asked, such as: When the sales volume (income) is high, average, and low, what are the gross profits and net income performances of the facility and for each profit center? How much difference in monthly income will be caused by changes in sales volume and what the manager will likely "see" on the control chart? When will a volume change occur and cause the service facility to enter the orange or red zone? Can a service manager predict how long the facility will be in the lower zones? Will a rebound into the green zone make up for the losses of a previous period?

With respect to costs and expenses, the volume analysis should ask similar questions. For example, do the variable costs (see Chapter 14) and expenses of the facility and profit centers vary with the changes in the sales (income) volumes? If so, how much (percentage) is the variation in each center? Are the changes immediate or is there a time lag? In general, a service manager must realize that a profit center with high variable costs should experience a lower proportion of losses when the volume of sales enters the orange or red zone, meaning the *proportion* decreases as income increases (see Chapter 14). However, for profit centers with high fixed costs that do not vary with sales and income (as discussed in Chapter 14), a larger proportion of income is needed to meet the costs as sales decline.

An analysis of costs and expenses must be conducted so that possible actions can be taken when sales and income decline to specific certain levels (such as turning down the heat or turning up the air conditioner, turning off lights, and closing a bay and reducing the working hours for hourly paid personnel, and so on). In other words, the volume of sales and their breakeven points based on the percentage of variable costs and fixed costs should be determined so that the contribution margin can be identified. A manager, therefore, could predict the breakeven for different sales volumes. Of interest to the manager, however, would be whether each profit center generates enough volume throughout the year (including the low-volume periods) to at least reach its breakeven point.

Financial Categories for Diagnosis

Financial data may be placed into different categories (listed below) for review. When collected, the information in the categories should be carefully reviewed to ensure the data is valid and accurate, and it will be used consistently from one period to another. For example, an advertising expense item for $1,000 could be entered in three time periods: when it was contracted (such as January), when the advertisement was run (say in February), or when it was paid (such as March). Accounting standards will recognize it one way but, for a management analysis, it should be recognized in February when the advertising actually ran and sales were hopefully stimulated. This would permit the manager to determine if the money spent on advertising resulted in a higher sales volume. Once a method is established (such as when advertising is recognized in a management analysis), it should be continued in order to note changes over time in the unit and volume analyses.

The categories that could be used in a unit analysis of profit centers and the facility include the following line items found on a traditional accounting statement:

- Specific sources of income
- Total income
- Cost of sales by line item
- Total cost of sales
- Gross profit
- Expenses line items
- Total expenses
- Net income

In addition, the volume analyses for a profit center and the entire facility could include the following categories obtained from nontraditional accounting records (although some information could in part be extracted from traditional accounting records):

- Number of sales (number of automobiles serviced)
- Flat-rate hours sold and available for sale
- Variable costs
- Total variable costs
- Contribution margin
- Fixed costs
- Total fixed costs
- Breakeven points

Comparing Data

When conducting a financial analysis, there are two basic methods used when comparing collected data. They should both be considered when using the above categories for profit centers as well as the entire facility. One is a **trend, or horizontal analysis**, in which data is compared from one year to the next (previous years are compared to the current year). The other method is a **static**, *or* **vertical analysis** in which data from one time period is compared to another time period within the fiscal year. The comparisons may be in the form of percentages (gross profit percentage of income) or dollars (amount of the gross profit), although, percentage comparisons are preferred.

In addition, a trend and static analysis may be used to compare the data of like units, such as the changes in gross profit of different profit centers. In such a comparison, however, the use of percentages versus dollars would be meaningful as the magnitude of the dollar amounts would be difficult to equate. A trend analysis study of the profit centers would see if the comparisons, such as the gross profit percentages, changed more for one center versus another from one year to another. A static analysis would look at changes in comparisons within the year (such as one month to the next). Therefore, the biggest difference between trend and static analysis is the time period. Trend examines data on a yearly basis, while static is used for a shorter period of time that is set by the user (daily, weekly, monthly, quarterly, etc.).

Examples of the Examination of Performances

Obviously, a discussion of all of the comparisons that could be conducted through the use of traditional and nontraditional financial data cannot be offered in this book because of the numerous possible combinations. This is even more complicated because each facility is different; such as type of facility, type of ownership, use of profit centers, location of the parts sales operation (in or out of the service unit), types and number of employees, and so on.

Consequently, the following subsections offer some case study examples of an analysis using the recommended process shown above and in the last chapter. The examples do not attempt to present a comprehensive analysis of a problem but focus on the process of stating a problem, asking a question, collecting of data, conducting an analysis, and discussing an option or options and conclusion.

Traditional and nontraditional financial data are shown below in Figure 15-1 for Renrag, Inc. The traditional information is shown as targets in a monthly income statement model; for example, the labor sales needed to be in the green zone is $50,000, along with nontraditional targets such as 1,000 labor hours at $50 per hour to make the income target. This model *does not attempt to exhibit an ideal or actual statement or a comprehensive review* but was designed for case study discussion purposes only.

Nontraditional Targets:
Labor Rate: $50 per hour.
Number of labor rate hours to be sold each month = 1,000.
Number of oil changes expected monthly = 200 @ $20.

RENRAG INC.
INCOME STATEMENT
Monthly Model

INCOME		
Labor Sales	50,000.00	59.5%
Sales – Parts and Accessories	30,000.00	35.7%
Sales Oil	4,000.00	4.8%
Total Income	84,000.00	100.0%
COST OF SALES		
Technician Wages	12,600.00	15.0%
Cost of Goods Sold – Oil	1,680.00	2.0%
Cost of Parts	18,000.00	21.4%
Total Cost of Sales	32,280.00	38.4%
Gross Profit	51,720.00	61.6%
Total Expenses	39,120.00	46.6%
Net Income	12,600.00	15.0%

FIGURE 15-1 Model statement for problems.

Problem: Labor Sales

Problem: Renrag's service manager reported to the owners that the facility's *labor sales had moved into the lower red zone* and would not earn the planned profit. The estimated income from labor sales was to be $50,000 (shown in Figure 15-1) to earn the net income of 15 percent needed to generate the annual planned profit. The actual labor sales were $38,000 for the month.

Question: Why did the labor sales move into the red zone?

Data Collected: Assume the service manager found that 1,000 flat-rate hours were sold (as per nontraditional target shown earlier). This included the labor sales sold on credit to local businesses.

Analysis: If 1,000 labor hours were sold and $38,000 was realized in labor income, the actual labor rate charged must be determined as follows:

Total Labor Sales/Total Flat-Rate Hours Sold = Actual Labor, Rate Charged.

$38,000/1,000 = $38 actual labor rate charged.

Conclusion: The actual average labor rate charged to customers was $38 per hour and not the posted labor rate of $50 per hour (shown in Figure 15-1). Therefore, the service manager must determine why there was a difference between the amount to be charged to customers and the average amount actually charged. To do so, the service manager must examine customer invoices to find out why less money was collected than expected. Possible reasons may be: basic mathematical errors, failure to use the stated labor rate on invoices, not enough labor hours were charged on invoices, awarding authorized and unauthorized discounts to fleets and other customers, and, of course, theft. Depending on the findings, the service manager may have to collect more information. For example, if mistakes were found on the invoices, the service manager must determine why and then either take action to correct the situation or make a recommendation to upper management or the owners.

Problem: Labor Sales

Problem: *Labor sales moved into the lower red zone* and the *planned profit will not be met* as expected by the owners. Again, the actual labor sales were $38,000 instead of the targeted $50,000.

Question: Why did labor sales move into the red zone?

Data Collected: Assume the service manager discovered that 760 FRHs were sold instead of the 1,000 FRH target.

Analysis: Because 760 flat-rate hours were sold instead of the 1,000 hours expected and the amount actually earned was $38,000, the amount charged per labor hour was $50 per FRH.

Conclusion: In this case, the conclusion would be that too few FRHs were sold, which then leads to another problem to be investigated (which is the next case study exercise discussed).

Problem: Technician Production and Efficiency

Problem: As per the previous case, the service consultant discovered that *the FRHs sold dropped to 760 FRHs* as compared to the 1,000 FRHs target (see Figure 15-1). This decline caused the labor sales to move into the red zone.

Question: Why were only 760 FRHs sold? The service manager must determine if each technician is pulling his/her load. This would lead to the following subquestions:

1. What was the efficiency performance for each technician? Did each technician reach his/her efficiency target?
2. What were the average number of flat-rate hours per invoice generated by each technician? How does this compare to the shop average?
3. What is each technician's job efficiency?
4. What does a trend and static analysis suggest with respect to FRH performances?

Data Collected: Assume that Renrag, Inc. has eight technicians who are expected to generate 1,000 hours per month. Because there is a mixture of A, B, C, and D techs, each has a different FHR target, or expectation, for the month. (Note: FRH targets should be used even if the technician is paid an hourly wage.) The service manager must collect all invoices (both cash and credit sales) and calculate the total number of flat-rate hours generated by the facility, each profit center, and each technician.

Analysis: In this case the analysis would address each subquestion separately.

1. What was the efficiency performance for each technician? Did each technician reach his/her efficiency target?

   The efficiency performance for each technician would be calculated by dividing the actual hours generated by a technician and by the technician's FRH target (the FRH expected to be generated by the technician). The formula for computing the efficiency rating for a technician would be as follows:

   > Technician weekly FRH earned / FRH target × 100 = FRH efficiency performance.

   For example, if a C technician (paid hourly) generated 30 hours for the week and the person's target expectation was 34 hours, the persons efficiency performance would be:

   > 30 FRHs generated/34 FRH target × 100% = 88% efficiency performance.

   In this case, the technician did not meet his/her efficiency target but fell short by 12 percent.

2. What were the average number of flat-rate hours per invoice generated by each technician? How does this compare to the shop average?

   The average number of flat-rate hours per invoice generated by each technician would be calculated by dividing the total number of flat-rate hours produced by the number of invoices as follows:

   > Technician FRHs/Number of Invoices = Average FRHs per Invoice.

   Assume that Ken, a technician, had an average FRH per invoice of 0.5 hours. Whether Ken's performance was good or bad can only be determined when it is compared to the shop average, which would be calculated by dividing the total shop FRHs sold by the number of invoices as follows:

   > Total shop FRHs Sold/Number of Invoices = Average shop FRHs per Invoice.

   If the shop average was found to be 2.1 FRH, it would indicate that Ken did not perform as well as the shop as a whole (he may just do the work on the repair order and perhaps not check over the automobile properly). Therefore, more information must be collected to examine Ken's problem in greater detail. Of course, if Ken is a D tech, he would likely have a low average because of the work assigned to him.

3. What is each technician's job efficiency?

Job efficiency is determined by dividing the number of hours taken by a technician to complete a job by the FRHs assigned to the job. In the formula:

N = Number of hours the technician punched on the repair order
FRH Assigned = Flat-rate hours assigned to the job.
FRH Assigned/N × 100 = Technician's efficiency per job.

Continuing this example for Ken, if his FRH efficiency (using the above calculation with question 1 above) was 60 percent and his FRHs per invoice was low (as per question 2), the service manager would then have to examine his efficiency on each repair order (question 3). Assume that in the review of Ken's job performances, the service manager found that Ken had a job that went really bad. Ken took 6 hours to make a repair that had .75 FRHs assigned to it for a job efficiency of 12.5 percent (.75/6). Naturally, more information must be collected to determine why he had such a low performance; such as, was he given a problem that was unfamiliar to him, is training or practice needed to improve his performance, is he not equipped properly to handle the assigned work, does he have personal problems that interfere with his work, and so on?

4. What does a trend and static analysis suggest with respect to FRH performances?

The service manager should go into more depth by conducting additional comparisons of the shop FRH average to past performances. A trend analysis, for example, may show that last year the shop also had a 2.1 FRH average per invoice, whereas a static analysis may indicate the number of FRHs produced had been improving over the past two months with an average of 2.4 FRHs per invoice. Further analysis must determine why there was a drop to last year's average. Is this simply a problem with a volume cycle in which sales will be low each year for this period? Could the poor performance of one technician (Ken) drop the entire shop average? To answer the first question, a trend analysis is needed for the past several years. To answer the second question, the potential increase in total sales would have to be calculated based on a 100 percent efficiency performance by Ken. If the amount of the sales comes close to reaching the target amount of $50,000, then Ken's performance is a problem.

Conclusion: The conclusions and recommendations to be made by the service manager would depend if all of the technicians were performing below their target levels or were off on only one week. If one week was off for all technicians, it could have been caused by a problem in the shop (such as a broken compressor or not enough customers) or due to a week when the shop was closed for one day for a national holiday. If FRH sales were off for the entire month and it is determined that it was not caused by a cycle in sales volumes or tech performance, the service manager must look at problems with the sale of labor hours (which is the next case study exercise).

However, before moving to the next exercise, assume that a technician, such as Ken, had an off month. The service manager must collect additional information beginning with an interview of the technician. In fact, when any technician has a low week, such as an efficiency performance of 20 percent, the service manager must interview the person. A low efficiency output by a technician can mathematically affect the service facility's financial performance. For example, in Ken's case, the service manager may find out that he had problems with the repair and that the customer should have been charged additional time for the repair or errors were made when the service consultant estimated the number of hours for the job. Naturally, the response and situation will dictate the follow-up questions the manager must ask in order to reach a decision on the action to take or recommendation to make to upper management and/or the owners.

Problem: Flat-Rate Hours Sold per Job

Problem: Only *760 FRHs were sold* and the review of the technicians' performances did not offer any conclusions.

Questions: To answer a question regarding the reason for the low sale of FRHs, the service consultant must answer the following:

1. How do the FRHs sold compare to previous periods?
2. Did the technicians recommend additional work to be sold based on their inspections of customer automobiles?
3. Based on the technician's inspection and recommendations for additional work, what percentage was the sales consultant able to sell?

Data Collection:

To answer these questions, the service manager must review repair orders (RO) to get the FRHs sold on each one when it was first given to the technician. The service manager should then record for each RO the additional FRHs recommended by the technician after the customer's automobile was inspected. Next the manager must enter for each RO how many of the technician's suggested FRHs were sold by the service consultant.

Analysis:

1. How do the FRHs sold compare to previous periods?

   Using the above information, the average FRHs sold should be compared to the same time period in previous years and to the time periods in the current year (static and trend analysis methods discussed earlier). This is to ensure that the facility was not experiencing a problem caused by a cycle of a low sales volume.

2. Did the technicians recommend additional work to be sold based on their inspections of customer automobiles?

   Using the above information to answer the above question and the next one, the three sets of data collected by the service manager would be represented by the following alphabetical letters.

   I = FRHs **i**nitially sold to the customer.
   P = FRHs of **p**otential sales the technician found as part of his/her inspection *excluding I.*
   S = FRHs actually **s**old or closed by the service consultant after the technician's inspection is completed *excluding I.*

   To answer the first question, the potential additional FRHs recommended by each technician would be divided by the FRHs entered on the RO when it was first received from the service consultant.

   P / I (X100) = % of increase in FRH reported by the technician.

   For example, if Randy was given an RO with 1.5 FRHs initially assigned to the job and returned the RO with suggested repairs that totaled 2.25 FRHs, he recommended a potential of an additional .75 FRHs for 50 percent additional potential sales (.75/1.5 = .5). After calculating all of the additional hours recommended and initially received for all invoices, an average number of hours recommended by each technician can provide an average percentage of up-sell recommendations.

3. Based on the technician's inspection and recommendations for additional work, what percentage was the sales consultant able to sell?

   To answer the third question, the potential hours less those that were sold by the service consultant would be divided by the potential hours recommended by the technician to gain the percentage of potential FRHs actually sold by the service consultant.

   (P − S) / P (X100) = % of FRH a service consultant actually sold.

To continue the example, the above repair order initially had an assigned 1.5 FRH (I = 1.5), which is excluded from S and P. After Randy looked over the automobile, he found an additional .75 FRHs of work (P = .75) to be done on it. Using the above equation, Bob, the service consultant, had an additional .75 FRH potential to up-sell. After calling the customer, assume Bob sold an additional .5 FRH (S = .5), which means he had up-sold 33 percent of Randy's additional recommended FRHs (.75 – .5) / .75 = .33 x 100 = 33%).

Conclusions: Customers come to a service facility and do not know what services to request. As a result, the initial repair order will often have a relatively low labor time on it, such as a maintenance, diagnosis, or inspection service. One of the jobs of the technicians is to thoroughly inspect the vehicle for potential problems. If problems are found, the service consultant must properly inform the customer (see Chapter 9 of text for service consultants by the authors).

Obviously, the conclusion depends on the reaction of the percentage of additional potential FRHs recommended by the technicians and then the percentage of up-sells by the service consultant. These expected percentages may appear as a business objective, such as an expected up-sell of 90 percent by the service consultant, which would create a target for a control chart. If the conclusion would be that the up-sells are below expectations and the performances in prior years or periods within the year, then an additional breakdown may be required, such as a division by profit centers or a division by the type of service provided (maintenance, diagnostic, or repair service). For example, one of the profit centers (such as the oil lube center or repair-diagnosis center) may be underperforming. There also may be reluctance on the part of a particular technician or technicians to examine vehicles carefully. It could be possible that a technician will examine only those systems on which he/she likes to work (as evidenced by the technician's average being far less than the shop's average). As the authors experienced in their business, some technicians cannot, or will not, examine automobiles for additional repairs or maintenance work. Unfortunately, this is a serious problem, for several reasons. One is that automobile facilities have a responsibility to their customers (who expect professionals to examine their vehicles) to make recommendations that will keep them safe from breakdowns. Second, if maintenance work is done and a problem not identified, the customer will likely go to another facility when the problem occurs because the service facility failed to find the problem. This causes a loss of confidence in the ability of the technicians at the facility. Furthermore, a decline in averages can be due to a change in the environment (as discussed in Part I of the book). For example, an environmental change may be an increase in the ownerships of new vehicles. This means the number of customers at an independent service facility may drop because of free maintenance and warranty repairs on new cars. In addition, if there is a question about the ability of the service consultant to close a sale (or unwillingness to sell additional repairs), an in-depth review is required via the collection of qualitative data (such as interviews and observations).

An additional note must be made here in regard to a review of FRHs in response to performances entering the upper orange or red zone indicating a sharp rise in a shop's sale of FRHs. This, of course, could be a result of working on older automobiles with more problems (if cars are too old, they often require more time. The additional FHRs also may point to an improvement in the technicians' ability to check over automobiles or a service consultant's skill in closing sales. For example, the most skilled service consultant employed by the authors was a former used-car salesman. At the same time, however, a service manager must be careful because the upper zone performances may point to the overselling of services and parts, especially if the service consultant is paid a commission. Taking advantage of customers in such a manner will cause unfortunate problems.

Problem: Gross Profit

Problem: The service manager found that Renrag's *gross profit fell to 52 percent* of total income (as opposed to the desired 61.6 percent shown in Figure 15-1).

Question: Did any of the line items in the cost of goods sold section of the income statement cause the grass profit to fall below the gross profit target of 60 percent?

Data Collected: The information needed would concern the cost of the parts and services sold and the gross profit generated from the sale of each of them.

Analysis: Each line item in the cost of goods section must be examined and compared to the income generated in the sale of that service or good to determine if the sale reached the targeted 60 percent gross profit. The calculation for this comparison would be to divide the cost of the part or service (such as labor or oil) by the amount of the income generated from its sale (such as labor or oil). The amount gained from the division is subtracted from one (1) to obtain the gross profit (multiplied by 100 percent to get the percentage of gross profit), which should be equal to or greater than (≥) the target. The formula for the calculation is:

$$1 - \text{Cost of part/Sale of the part (X 100)} \geq \text{gross profit target.}$$

Assume the cost of parts for Renrag was $24,000. In this case the gross profit on the sale of parts would be:

$$1 - \$24{,}000/\$30{,}000 = 20\% < 60\%.$$

Conclusion: Assume from the calculation of the parts sales, the gross profit earned from their sale was 20 percent and less than (<)60 percent. The service consultant would have to check the markups on all parts sold to see if errors in markups were made, if the facility had been charged for parts it did not sell, if a large inventory delivery was charged to parts instead of going into an inventory account, or if someone or some people were not charged for their parts. All of these question, plus numerous others, could be reasons for the poor statistic.

Problem: Expenses

Problem: The income statement for Renrag indicated the net income dropped 4 percent (to 11 percent) even though there was an increase of 3 percent in gross profit (to 64.6 percent). On checking, the service manager found that *total expenses increased 7 percent* to 54.6 percent.

Question: What caused the increase in expenses?

Data Collected: A manager's job is to examine expenses each month to make sure they are within a specified range. This requires the manager to look at each expense line item for every profit center and compare it to a target metric. The target metric should be shown on the worksheet for each profit center but, in reality, may change from season to season and even from month to month. For example, heat, air conditioning, and electricity are dependent on the weather. Although the target metric needs to take these changes into consideration, there may be exceptional circumstances that caused a spike in one of the line item accounts.

Analysis: The service manager must determine if a spike in an expense line item or line items exists. Assume that a spike was not found in any line items. As a result, the service manager must examine all of the expense accounts more carefully to determine if any of them increased and the percentage for each. In this case, the changes in a number of accounts would probably be greater than normal and add up to 7 percent.

Conclusion: Assume the service manager's examination of all of the accounts found that several had nominal increases. The service manager must examine each one to ensure they were justified and check the target metrics for accuracy. In some cases, the changes may be temporary and will return to normal. If not, then the target metric may need revised and labor rates and/or parts markup may need to be increased.

Employee Theft

This chapter on the analysis of financial performance cannot conclude without addressing the problem of theft. As discussed earlier, one of the important jobs of the service manager is to ensure that theft does not occur. There are many ways to prevent the theft of major items or continuous theft of smaller items, but there is an old rule that a business should not spend more money guarding resources than what the resources are worth. For example, a business should not put in an alarm system that cost $2,000 to guard a room that never contains more than $50.

On the other hand, there are numerous inexpensive practices that can be used to protect company resources. For instance, the use of sequentially numbered repair orders/invoices is inexpensive. If a service consultant makes an error on a repair order/invoice, it must be voided and attached to the copy of the new invoice for submission. Therefore, the assumption is that if an invoice is missing, then money is missing, meaning that a customer paid cash for a service and an employee put the money in his/her pocket. The service manager must account for all numbered invoices. When an invoice is missing, the service manager should immediately (not next week) search for the invoice. If it cannot be found, then an audit of technician time tickets to match jobs to invoices is required. In addition, deposit slips must be checked to determine how much money was collected and deposited. The daily deposits must match the cash payments received for the day. The amount left in the cash drawer at the end of the day must equal the cash in the drawer when the business day began plus the amount of cash collected minus the deposit. The service manager must monitor the handling of cash and the cash drawer everyday. Another security procedure is to keep the credit card machine in a restricted place during the day and preferably locked up at the end of the business day. At one company, after the business day ended, some employees were using the company credit card machine to run credit charges for their personal credit cards (reducing the balance they owned to the credit card company against their employer's account).

Unfortunately, the authors can testify that employee theft occurs more often and in more ways than one can imagine. In many of those cases, thefts occur because managers are careless while in others, the thief goes to great lengths to take money or things that do not belong to him or her. When unchecked, employee theft is the greatest threat to a company's resources, even when including theft by people who break into a facility after hours. Security checks of a company's resources must be made on a regular basis for two reasons. One is to ensure the resources are there. The second is to show all employees that the security checks are conducted.

Some managers and owners may believe that computers and computer programs can serve to protect a company's resources. Although this may make it easier to monitor company resources, computer programs can easily be manipulated and modified. For example, a temporary increase in the labor rate can be as easy as a few keystrokes and at the end of the business day, the extra cash can be skimmed from the top and the books will appear normal. In these cases, few managers would check to see if the labor rate for a each job was off by $.20 to several dollars per hour, yet it may be enough to skim a few hundred dollars a week off the top of cash sales. Consequently, service managers cannot expect computers to offer any security for accounts, customers, correspondence, purchases, sales, and so on; rather, managers must constantly monitor the computer operations by becoming an expert on how the programs work. The bottom line is that if an employee is discovered stealing anything from the company, the person must be terminated and possibly the theft reported to the appropriate law enforcement authorities.

Final Thoughts

There is no question that the procedures and processes described in this book are not simple or easy. Some owners and managers may question the need to use these techniques or may want to use other techniques not discussed. However, the changes in business over the past 30 years make their use more necessary than ever, especially when the difference between profit and loss may come down to a few critical decisions.

At one time, the owners and managers of a facility could fly by the seat of their pants, much like the pilot of an old biplane. Today, the management of a business (and the flying of airplanes) cannot continue get by with this practice of the past. While some of the fundamentals may still hold true, the reality is that technology and the use of computers has made the process more complex than ever before. Machines and computers have made life easier but they also permit human beings to do much more than they could 30 years ago. From a management perspective, it is great that one person can do the work of two or more people so that the business can handle more customers and earn more money. With this growth, however, comes more expenses that drive up costs. For instance, a small automotive facility with a computer and one service consultant may be able to schedule 50 customers a day for service as well as order parts, prepare estimates, print invoices, collect payments, maintain service records, and contact customers to inform them of their next service. The use of computers and software programs has dramatically changed the job of everyone at a service facility, including the service manager.

In addition to the changes in the operations of the automobile service business, owners and managers have had to cope with the incredible changes that have taken place in the automobile and with its customers. At one time, many automobile owners serviced and fixed their cars. Being able to change the plugs, set the timing, change shocks or brakes, and take care of the cooling system was not a big deal. Cars were simple machines. For example, 50 years ago a technician, called a mechanic then, by the name of John owned a one-bay garage. He was the service consultant, technician, bookkeeper, and janitor. Operating a business then was quite simple and customer expectations were different. One day, a customer came to his shop and told John to replace the spark plugs in his car. At the end of the day, the customer picked up his car and left. In 15 minutes, he returned and told John he did not fix his car. John responded that the customer did not ask him to fix it but simply to change the plugs. Although this may seem strange today, it was not unusual at John's facility. Customers would ask for a particular service, such as changing the plugs, and then to save money would finish the job at home (if anything else was needed). Possibly, the owner improperly diagnosed the problem and thought his car just needed plugs. Today, this case would be as unlikely as a one-man service facility. (Of course, the authors had customers bring their automobiles to the shop in parts because they could not put it together again.) Owners of today's complex automobile hardly attempt to fix them. Instead, they depend heavily on their service facility to keep their cars on the road.

Consequently, the operations of a service facility and the repair of automobiles call for professional leadership. The position of the service manager has evolved to occupy a professional status, meaning it requires special learning and carries a certain prestige. Service managers must supervise and direct the daily operations of the service facility, including personnel and financial activities. The financial security of a facility and its employees depend heavily on the special knowledge and ability of the service manager. The expectations of the owners of the facility and their corresponding dependability on service managers have increased in recent years. Although this book has covered the basic content needed by service managers to fill this professional role, there are more advanced practices and techniques to be learned. In addition, as business techniques and automobiles continue to advance, so will the tasks, duties, and expectations of service managers.

Review Questions

1. What process is to be followed to diagnose a facility when performances are recorded in the lower orange and red zones?
2. What is the difference between traditional and nontraditional financial data?
3. What is the purpose of a unit analysis and a volume analysis?
4. What is the difference between a trend and static analysis?
5. What steps are to be followed when conducting an in-depth review of a performance problem?
6. Outline a plan to maintain a collection of traditional and nontraditional financial information for a service facility?

PART IV

PRACTICAL EXERCISE

Small Group Breakout Exercises

1. As explained in an earlier practicum, the owner has set a goal that called for an annual 10 percent increase in the sale of services. Last year, the increase in sales for the service department was 7.5 percent. In addition, the financial statement for June 30 indicates the increase may be below 6 percent at the end of the current fiscal year. The owners have requested that you explain how you will analyze the service department and your profit centers to describe why the 10 percent goal was not been reached.
   A. To begin this assignment, assume that two years ago the service department sold $650,000 in labor, which included 13,000 hours from five technicians servicing 9,000 vehicles. The past year, the increase in 7.5 percent included an increase in sales to $698,750 but the 13,000 hours from the five technicians remained the same for the 9,000 vehicles serviced. Prepare a report that describes how you will conduct a unit analysis of the service department's performance (sale per vehicle) and the technician performance (labor hours per vehicle and labor hours per technician). Generate a report that not only thinks in terms of the sales

but also in terms of other measures that will help you analyze and reach your targets.

B. Create a service department profit center to describe the cost to deliver a particular service (for example, create a tire replacement profit center). Use a labor guide to obtain accurate times for the profit center and equipment catalog to obtain equipment costs. Assume your technicians are all paid by flat rate and the C technicians cost $10 per hour, B technicians cost $15 per hour, and A technicians cost $20 per hour. Assume your overhead is a fixed cost of $0.1 per square foot per hour or stated another way, $30 per bay per hour (this is a working area space of 12 foot by 25 foot space or 300 square feet). The labor rate is $60 per hour. Determine whether you can deliver the services at a competitive price and can also recover the equipment cost. For example, lower-skilled workers will help to cut cost (provided quality services can be delivered), while expensive equipment will add cost. Depending on how you set up the analysis, you may find that an additional charge must be added into the price of the service to cover the cost of the equipment. In this exercise, the calculation of your contribution margin and determining breakeven for the profit center you choose to create and analyze is key.

C. Prepare a report that describes how you will conduct a volume analysis of the service department. Think in terms of how many cars must be serviced (invoices processed) or how many labor hours sold to reach your goals and/or breakeven for the service department or a particular profit center.

D. Include with your reports an example of a set of control charts that you will use to visually present your ideas. Specifically the control charges should show how income is related to costs, and profit of the service department andor profit center.

2. You have been given a memo from the state directing all service facilities to limit the charges for the new state emissions test to $9.00. Their rationale for this charge is that the test takes twelve minutes, or two-tenths (0.2) of an hour, to conduct the emission test. The state administrator bases the charge of $9 on a technician who earns $22.12 per hour (state average according to the census bureau). In addition, they also assume that each facility will conduct 1,800 inspections over three years (600 per year) and this will be enough to generate enough money to pay for the $8,000 each service facility must invest to purchase the new emission machine. As an example, they sent you the following breakdown for the $9 charge:

Technician's labor = $4.42 (.2 × $22.12)
Emission cost recovery = $4.58 ($8000/1800)
Total $9.00

Your problem, however, is that your hourly bay rate is $40. This rate is necessary in order to cover the technician's labor plus overhead (such as heat, light, electricity, service consultant's time, etc.) and for a profit earned for the owners. The owners tell you that they feel the equipment cost should be treaded like a three-year loan at 15 percent interest (cost recovery of the capital is within three years) and you must earn $278 per month, or $3,336 per year, to pay for the equipment. You can calculate that based on 600 emission inspections (18,00/3 years = 600 per year) requires a receipt of $5.56 per inspection ($3,336/600).

Based on this information, you must cover $3,336 for equipment charges. You also must cover your technician labor and overhead plus make a profit. This means you must earn $8 per inspection (two-tenths per hour at your labor rate is .2 × $40 per hour = $8). Unfortunately, the current emission program (which basically tests gas caps) does not provide opportunities to up-sell (earn money from making needed additional repairs) and going against the state's recommended fee can be a public-perception problem.

A. Calculate how many inspections you must do each year at $9 to break even (cover labor/overhead/profit at $8 plus earn enough to pay for the equipment).

Next, create a control chart with your targeted breakeven (number of inspections you calculated × $9) and a lower control limit of $5,400 ($9 × 600 inspections estimated by the state). Then do the math (at $9 per inspection) and plot the points (on the horizontal axis of the control chart) for the following emission inspections each year: 600, 1,100, 1,600, 2,100, 2,600, 3,100, and 3,336. Answer the following questions:

B. How far (in dollars) are you below the breakeven at each interval?

C. How far (in percent) are you below the breakeven at each interval?

D. If the state is correct and your dealership will only do 600 inspections, what will this mean to the profit of the business? To do this, take how much is owed for the equipment each year ($3,336; given above) and subtract the following: How much you earned to pay for the equipment per inspection times 600 inspections.

E. If you make your breakeven target, you will have covered your labor costs, overhead, and profit for each inspection plus earned enough to pay for the equipment each year. If you exceed the number of inspections needed to break even, the extra money earned will add to your profits (as the equipment has already been paid for). How much profit can you expect if you exceed the number of inspections needed to break even by 500 inspections?

F. If the owner's goal is to have a 10 percent increase in sales (which is currently $750,000 per year), will this program help you achieve this goal if you achieve breakeven plus 500 more inspections? How much more must you earn in addition to the inspections to reach the 10 percent goal?

GLOSSARY

Accounting—a process that records, classifies, and summarizes business transactions.

Accrual basis of accounting—a sale is entered at the time the sale is made (whether or not cash is received) and expenses are entered at the time of the purchase (not when cash is paid).

Active delivery—when a service consultant personally delivers an automobile to a customer to discuss matters of importance, to check the condition of the automobile, and to determine customer satisfaction.

Appraisal—evaluation of employee performances in reference to the tasks in the person's job description.

Assets—accounts in financial records that indicate what a business owns.

Automobile service facility—a profit business that performs the maintenance, repair, and diagnosis on automobiles.

Balance sheet—a financial statement that presents a company's assets, liabilities, and owner's equity as of a specific date.

Benchmark—a specific date representing the end of a period of time, such as a week, month, quarter, fiscal year, when a review will be conducted to determine if an objective (target) has been achieved.

Budget—the allocation of a specific amount of money, usually to expense categories in which it is to be spent, to an operating unit, or profit center, in a business.

Bumper-to-bumper warranty—a contract that covers all of the components in an automobile (see also *warranty contract*).

Business transaction—when money or something of value is exchanged for goods or service.

Campaign repairs—see *recall repairs.*

Capital assets—fixed assets, such as buildings, land, and equipment.

Cash basis accounting—a sale is recorded when cash (payment) is received and expenses are entered when cash is paid.

Cash flow statement—presents the flows of money into and out of a business to examine the change in cash for a given a period of time.

C Corporation—a corporations taxed under subsection C of the Internal Revenue Service (IRS) code.

Chain service facility—an automobile service facility that is one of several facilities owned by a corporation, such as Pep Boys and Sears, Roebuck, and Co.

Closing a sale—when a customer gives an approval for a service by signing a repair.

Collective bargaining unit—a group of employees (called union members) who negotiate their wages, benefits, and working conditions as a group with the owners/management of the service facility.

Comebacks—occur when a customer had a repair made to an automobile and must return to have the same repair made again.

Company policies—statements that indicate how the owners want to conduct business and legal regulations that direct the way business and services must be conducted.

Cores—a used part that has been replaced and must be returned to the supplier for a reduction in the cost of the new part.

Corporate bylaws—rules and regulations that govern the corporation including provisions regarding the election by the shareholders of the corporate directors.

Corporate directors—often referred to as the Board of Directors or Corporate Board, represent the interests of shareholders, appoint and supervise the performances of corporate officers, and oversee general corporate business matters.

Corporate officers—the president, vice president, treasurer, and secretary of a corporation.

Corporation—a business that becomes a legal entity owned by one or more people who invest in the business by purchasing shares of stock.

Costs—selected expenses incurred for the delivery of a specific service or in the sale of a specific product.

Criterion measure—the comparison of a measure to a standard.

Cross-training—when workers are taught how to perform other jobs in a business.

Current assets—cash or an asset that can be converted to cash in a short period of time without disrupting business operations.

Current liabilities—short-term credit arrangements to be paid within the fiscal year.

Customer automobile inventory sheet—a form used to record customer automobiles located in the building and on the property of the facility.

Customer status sheet—a form used to record customer automobiles being serviced and the status of their services from the creation of the repair order to the payment of the invoice.

Daily customer log—a calendar with pages to enter daily appointments, vital information on each customer, services to be performed, and the estimated time needed for each service.

Dealership service facilities—facilities that concentrate on the performance of maintenance, repair, and diagnostic services on specific makes and models of new and used automobiles.

Depreciation—the allocation of the cost of a capital asset (except land) over its expected useful life.

Development—keeping employees up-to-date on the tasks and duties required in their job.

Dividend—the portion of the profit earned by a corporation that is distributed to shareholders.

Effectiveness—when work is done correctly and quality outcomes are achieved.

Efficiency—when resources are used properly and output is maximized.

Emission warranty—a federal requirement that new-automobile manufacturers must guarantee the repair of their automobile emission components for a stated period of time.

Employer Identification Number (EIN)—a business identification number obtained from the federal government.

Enterprise resource planning system (ERP)—uses information in traditional accounting statements and reports, as well as nontraditional accounting information, to make management decisions.

Expenditure—the spending of revenue awarded to a business or government body.

Expense—the spending of income earned by a business.

Expense transaction—when a facility pays money for goods or services.

Extended warranty—a warranty that is purchased by owners of new and used automobiles to provide warranty coverage (sometimes with a deductible) for identified components in an automobile for a given period of time or number of miles the automobile is driven from the date of purchase.

Feature-benefit selling—a sales approach that explains the service to be provide (feature) and the advantage (benefit).

Federal Insurance Contribution Act (FICA)—Social Security/Medicare taxes paid by employees and employers.

Financial accounting statements—report the present financial position of a business.

Fiscal year—the end of a 12-month business year.

Flat-rate objective—the number of flat-rate hours a technician hopes to work in a week.

Flat-rate pay—wages paid for a specified amount of time that it takes to perform a job.

Fleet service facility—a service facility that limits its services to vehicles owned by a company or government body.

Flow diagram—a drawing that illustrates the processing of work or an activity through a business or organization.

Formative appraisal (evaluation)—a narrow, focused, confidential assessment designed to improve performance.

Formatted system—when an operations manual is followed to encourage consistency in performances of an automotive repair facility.

Franchise—a business granted the use of a nationally recognized name in return for a fee and a percentage of sales.

Garage keeper's policy—liability insurance that covers damages to customer automobiles while at a facility.

General independent service facility—a facility that performs maintenance, repair, and diagnostic work on all models and makes of automobiles.

Goals—visionary and idealistic statements declaring the aspirations of the company that support the company's mission.

Gross income—the total amount of money earned when earned revenue is added to gross sales received from the beginning to the end of a business period (day, week, month, year).

Gross profit—the balance left over from gross sales after subtracting technician salaries and the cost of the parts used to repair customer automobiles.

Gross sales—the total amount of money received from the sale of goods or services less any sales returns.

Hard copy (technician's)—a thicker, cardboard-like copy of a repair order given to the technician.

Income—money earned from the sale of goods and services.

Income statement—presents the amount of money earned by a business minus the expenses to show the profit or loss for over a specified period of time.

Induction—to assist new employees in getting accustomed to his/her new position and become familiar with the way the company conducts operations.

Job classification—a rank or standing that is relevant to a position, such as an A tech.

Job description—a list or description of the job tasks to be performed by a person in a position.

Job duties—details of job tasks.

Job rank—signifies the authority or responsibility assigned to a position, such as manager.

Job tasks—major work assignments to be conducted by a person in a specific job.

Labor sales worksheet—designed to record information on labor sales, cost of labor sales, and related expenses.

Lead technician—a technician who assists with the coordination of the work in a shop, in assigning jobs to other technicians, in quality control, in monitoring the condition and use of equipment, and with other supervisory duties assigned by management.

Lemon laws—laws that require automobile manufacturers, and in some states the automobile dealers, to buy back an automobile when it is not properly repaired.

Liabilities—accounts in financial records that indicate what a business owes.

Line manager—an employee who has the authority to make company decisions and set company policy, such as a change in the labor rate to be charged to a special account.

Liquidation—when a business uses its cash and sells everything it owns to pay its debts.

Limited Liability Company (LLC)—a legal entity such as a corporation; however, the owners are referred to as members (a person or another entity like a corporation) and make the business decisions for the company unless they specify otherwise in their articles of organization.

Long-term assets—property, equipment, and other assets that cannot be converted to cash in a short period of time, such as a year.

Long-term liabilities—credit arrangements to be paid in one year or longer.

Maintenance contract—awarded to or purchased by a customer at the time an automobile is purchased; it pays for specified maintenance services for a set period of time from the date of purchase or set number of miles an automobile is driven.

Markup—the difference between the amount a facility pays for a part and the amount it is sold to a customer.

Mechanics lien—state laws that permit a service facility to hold a customer's automobile until payment for a service is received.

Mission—states why a business exists and its purpose.

Net loss—a negative balance after expenses and business taxes are deducted from the gross profit.

Net profit—a positive balance after expenses and business taxes are deducted from the gross profit.

New-automobile manufacturer warranty—a contract awarded to the owners of an automobile that provides for the repair of the automobile at no charge for a predetermined length of time from its initial purchase or for a predetermined number of miles that the automobile is driven.

Nonexempt employees—job classifications as per by the federal government in the Fair Labor Standards Act (FLSA) to indicate a position is not exempt from the hourly regulations of the act.

Nontraditional accounting information—quantitative information, not typically found in traditional accounting statements and used by managers to gain insights into business operations to make management decisions.

Normative measure—the comparison of a measure to other measurements.

Objectives—a set of statements detailing what will be accomplished.

Open business environment—forces outside the control of a service facility that affect its sales potential.

Operational environment—features or forces that have an influence on the daily business activities of a service facility.

Operational plan—a plan that describes how a business will function or work to meet the expectations presented in the mission statements, goals, and objectives.

Operational procedures—procedures that present objectives and performance expectations for each operation in a company.

Operation manual—a manual that directs the way work is to be conducted and processed in each work area.

Operations research—the examination of processes, such as procedures followed by a service facility to make repairs on automobiles.

Organizational diagram—a diagram that presents the organizational structure of a company or business.

Organizational structure—the managerial chain of command and the relationships among the positions in the chain of command.

Overhead—indirect expenses incurred in support of the delivery of services and sale of goods.

Overselling—when customers are sold a service their automobile does not need.

Owner's equity—represents the amount of money a business is worth after liabilities are subtracted from the assets.

Partnership—a business owned by two or more people.

Petty cash account—a fund that makes cash available for the purchase of items below a specified amount of money.

Policy check—a voucher given to customers for the purchase of goods or services from the service facility.

Prepriced maintenance menu—a chart that presents different maintenances suggested by a manufacturer and the charges for each.

Product specific service facility—a service facility that diagnoses, repairs, and performs maintenance on specific makes and models of automobiles, such as a facility that services only Volkswagens.

Profit—when income exceeds expenses.

Proprietorship—a sole owner who provides the money needed to open, buy equipment, and maintain initial operations of a business.

Qualitative information—information about quality from observations, insights, judgments, and perceptions, such as customer opinions about the quality of repairs and services.

Quantitative information—information gained from measurements.

Recall (campaign) repairs—repairs undertaken when a manufacturer requests the owners of a specific year, make, and model of automobile to take it to a service facility for repair at no charge to the owner.

Recruitment—attraction of qualified people to apply for a position.

Repair categories—a list used to indicate the relative importance of a suggested repair.

Repair order–tracking sheet—a form used to monitor the status of each repair order.

Revenue—money received from a government body for goods or services to be delivered and from investments, such as interest earned or dividend paid. In some textbooks, the term revenue is used interchangeably with the term income.

Revenue transaction—money received from sources, such as interest received from loaning money to a bank through a savings or checking account.

Sales transaction—money received from customers from the sale of goods or services.

S Corporation—a corporation that is taxed under subchapter S of the Internal Revenue Service (IRS) code, which permits it to avoid paying taxes on earnings.

Seamless system—a process that is not disrupted when work is passed from one person or one station to another.

Selection—hiring the best-qualified candidate for a job.

Service—the maintenance, repair, and diagnosis of an automobile.

Shaping process—the reinforcement of acceptable performance and redirection of performance deficiencies.

Shop management system—the procedures, documents, files, and computer used to prepare, store, and retrieve customer information, repair orders, automobiles serviced, flat-rate hours, parts markups, parts inventory, and vendor information.

Span of control—the number of employees a manager personally supervises.

Specialty service facility—a facility that provides services to specific makes and models of new or used automobiles, such as a dealership.

Specialty team—a team whose repairs, maintenance, or diagnostic work assignments are limited, such as a team that performs only maintenance work.

Staff exempt employees—job classifications as per by the federal government in the Fair Labor Standards Act (FLSA) to indicate a position is exempt from hourly regulations in the act.

Staff manager—an employee who provides support to line managers and, as a result, is assigned duties by a line manager.

Static or vertical analysis—data from one time period that is compared to another time period within a fiscal year.

Stockholders—people who invest in a corporation by purchasing shares of stock in the business.

Stockholder's equity—the amount of money the owners or stockholders would have if the assets were liquidated to pay off all liabilities.

Strategic business plan—presents the strategic, tactical, and operational plans of the company with an organizational chart, management team, targets, marketing plan, description of facility, financial information including income projections, targets (expectations) with benchmarks, and contingency plans.

Strategic plan—the mission, goals, and objectives of the company.

Sublet sales—when a facility sends a customer's automobile to another facility for service, has the automobile returned when completed, pays the other facility for the service, and charges the customer for the service conducted by the other facility.

Summative appraisal evaluation—a broad, formal evaluation used for decision-making purposes, such as a promotion, pay increase, suspension, or termination.

System specific service facility—a service facility that repairs and maintains one automobile system, such as transmissions or brakes.

Tactical environment—factors that influence the provision and supply of resources needed to conduct business activities.

Tactical plan—a plan that identifies resources needed by management to meet company objectives

Target—a proposed objective to be obtained at the end of a specific time period.

Target metric—optimum operating tolerances (numerical ranges, specific numbers, or percentages) that each line item on the worksheet should operate within.

Technical service bulletin (TSB)—announcements put out by automobile manufacturers to dealerships and subscribers of computer repair information systems to announce automobile operational concerns and how to repair them.

Trend or horizontal analysis—when data is compared from one time period, such as a year, to the next (previous years are compared to the current year).

Up-selling—when a customer selects higher-quality parts or labor workmanship as opposed to less-expensive parts or labor.

Value-added service—a service received by a customer at no extra cost.

Vehicle identification number (VIN)—a unique 17-digit number assigned to an automobile when it is manufactured.

Vendor—a company that sells parts, goods, and services to a facility.

Warranty contract—a document that specifies that if a part does not work properly or breaks within a set period of time, a new part, and possibly the cost of the labor to replace it, will be provided at no charge to the customer.

Work flow—the processing of work from the initial contact with customers to the return of their automobile.

Workman's compensation—an insurance policy covering injuries received by employees when at work.

INDEX

O

P

Q

R

S